PIC Robotics

PIC Robotics

A Beginner's Guide to
Robotics Projects
Using the PICmicro

John Iovine

McGraw-Hill

Francisco Lisbon London Madrid
Milan New Delhi San Juan Seoul
Singapore Sydney Toronto

The *McGraw·Hill* Companies

CIP Data is on file with the Library of Congress

1 2 3 4 5 6 7 8 9 0   DOC/DOC   0 9 8 7 6 5 4 3

ISBN 0-07-137324-1

The sponsoring editor for this book was Judy Bass, the editing supervisor was Stephen M. Smith, and the production supervisor was Pamela A. Pelton. It was set in Century Schoolbook by Paul Scozzari of McGraw-Hill Professional's Hightstown, N.J., composition unit. The art director for the cover was Margaret Webster-Shapiro.

Printed and bound by RR Donnelley.

McGraw-Hill books are available at special quantity discounts to use as premiums and sales promotions, or for use in corporate training programs. For more information, please write to the Director of Special Sales, McGraw-Hill Professional, Two Penn Plaza, New York, NY 10121-2298. Or contact your local bookstore.

This book is printed on recycled, acid-free paper containing a minimum of 50% recycled, de-inked fiber.

Contents

Preface

This is a project book on building small robots. Each robot utilizes the PICmicro series of microcontrollers from Microchip Technologies Inc. for intelligence, navigation, motor control, and sensory readings. By changing the microcontroller programming and sensory electronics we can create a zoo of robots that includes photovores, behavior-based (neural) robots, hexapod and bipedal walkers, and artificial vision systems that can track and follow objects.

Each robot project has something to teach.

John Iovine

Robot Intelligence

The robotic projects outlined in this book make extensive use of the PIC series of microcontroller from Microchip Technology Inc. In addition to its ability to run programs, the microcontroller has input and output lines (pins) that are used to control motor drive systems, read sensors, and communicate. We demand a lot from our microcontroller(s), so it's important to have a good idea of what a microcontroller is right from the start.

What Is a Microcontroller?

A microcontroller is essentially an inexpensive single-chip computer. Single chip means the entire computer system lies within the confines of a sliver of silicon encapsulated inside the plastic housing of an integrated circuit. The microcontroller has features similar to those of a standard personal computer. The microcontroller contains a CPU (central processing unit), RAM (random access memory), ROM (read-only memory), I/O (input/output) lines, serial and parallel ports, timers, and sometimes other built-in peripherals such as analog-to-digital (A/D) and digital-to-analog (D/A) converters. The key feature, however, is the microcontroller's capability of uploading, storing, and running a program.

Why Use a Microcontroller?

Being inexpensive single-chip computers, microcontrollers are easy to embed into larger electronic circuit designs. Their ability to store and run unique programs makes them extremely versatile. For instance, one can program a microcontroller to make decisions and perform functions based on situations (I/O line logic) and events. The math and logic functions allow the microcontroller to mimic sophisticated logic and electronic circuits.

Programs can also make the microcontroller behave as a neural network and/or a fuzzy logic controller. Microcontrollers are incorporated in consumer electronics and are responsible for the "intelligence" in these smart electronic devices.

Designer Computers—So Many Microcontrollers

There are a large variety of microcontrollers on the market. We will use the versatile microcontroller chips called PIC chips (or PICmicro chips) from Microchip Technology Inc.

The Compiler

There are a number of compilers on the market that allow users to write programs (code) in different high-level languages. High-level language frees the programmer from wrestling with and controlling the microcontroller's registers when writing code and accessing the different aspects of the microcontroller's features and memory.

The high-level language I use is a derivative of the Basic language. It is called PicBasic. (The PicBasic and PicBasic Pro compilers used to write PicBasic programs are products and trademarks of microEngineering Labs, Inc.) PicBasic is similar to the PBasic language used in programming the Basic Stamp series. Programming microcontrollers directly using the PicBasic (or PicBasic Pro) compiler offer two major advantages over the Basic Stamp series of microcontrollers which use external serial EEPROM for memory storage, faster program execution speed (20- to 100-fold increase), and reduced cost.

PIC Programming Overview

Programming PIC microcontrollers is a simple three-step process: Write the code, compile the code, and upload the code into a microcontroller. Following is an overview of the process; step-by-step instructions will be provided in the following chapters.

Software and Hardware

You will need two items to begin programming and building microcontroller-based projects and robotics. First is the compiler, either the PicBasic Pro or PicBasic compiler (see Fig. 1.1). The PicBasic Pro compiler from microEngineering Labs, Inc. has a suggested retail price of $249.95. The PicBasic compiler from microEngineering Labs, Inc. has a suggested retail price of $99.95. In addition to a compiler you need the EPIC programming board and software; this package sells for $59.95 (see Fig. 1.2). (EPIC is a product and trademark of microEngineering Labs, Inc.)

Figure 1.1 PicBasic Pro and PicBasic software packages and manuals.

Figure 1.2 EPIC Programmer software and hardware.

PicBasic and PicBasic Pro Compilers

The PicBasic and PicBasic Pro compilers both function in the same way. Saved program code (text file) is run through a compiler (either the PicBasic or PicBasic Pro compiler). The compiler reads through the text file and creates (compiles) an equivalent machine code instruction listing (.hex file) of the program. The machine code (.hex file) is a list of hexadecimal numbers that represent the PicBasic program. The list of hexadecimal numbers (.hex file) is uploaded (programmed) into the microcontroller. When the microcontroller is started, its CPU will run through the programmed list of hexadecimal numbers, running the PicBasic program. Uploading the machine code (.hex file) into the microcontroller is the job of the EPIC Programmer board and software, which we will look at shortly.

The PicBasic Pro compiler is considerably more expensive than the standard PicBasic compiler. The Pro version offers an enhanced and richer basic command syntax than is available in the PicBasic compiler package. A few of the additional commands that can be found in the Pro version allow the use of interrupts, direct control of LCD modules, DTMF out, and X-10 commands, to name a few.

While the PicBasic Pro is a more sophisticated package, the compiler does not handle two of my favorite Basic commands, peek and poke. Although the commands are listed as "functional" in the Pro manual, it is emphasized that "PEEK and POKE should never be used in a PicBasic Pro program." There are work-arounds to using the peek and poke commands in the Pro version that will be covered when needed later on.

In the balance of this book, at times I will refer to both the PicBasic and PicBasic Pro compilers simply as the compiler(s). This saves me from continually writing PicBasic and PicBasic Pro compiler throughout the book. When a distinction becomes necessary, I will specify the individual compiler.

The compiler program may be run manually in DOS or in an "MS-DOS Prompt" window. A third option, and one you will probably use, is to run the compiler within a Windows program called CodeDesigner. CodeDesigner is discussed later in this chapter and fully in Chap. 4.

The minimum system requirement for the compiler is an XT-class personal computer (PC) running DOS 3.3 or higher. The compiler can compile programs for a large variety of PIC microcontrollers.

EPIC Programmer

The second item needed is the EPIC Programmer, also made by microEngineering Labs, Inc. The EPIC Programmer consists of software (EPIC) and a programming carrier board (hardware). The EPIC software package has two executable files, one for DOS and another version of the software for Windows.

It is the EPIC hardware and software that takes the compiled .hex file generated by the compiler and uploads it into the microcontroller, where it may be run. The EPIC Programmer is compatible with both the PicBasic and PicBasic Pro compilers.

The programming carrier board (see Fig. 1.3) has a socket for inserting the PIC chip and connecting it to the computer, via the printer port, for programming. The programming board connects to the computer's printer port via a DB25 cable. If the computer only has one printer port with a printer connected to it, the printer must be temporarily disconnected to program PIC chips. The EPIC programming carrier board supports a large variety of PIC microcontrollers.

Figure 1.3 Close-up of EPIC programming carrier board.

Firmware

Many writers use the term *firmware.* This word is used when software is embedded in a hardware device that can read and execute by the device but cannot be modified. So when our program (software) is embedded (uploaded) into the microcontroller, it may be referred to as firmware. Other phrases may include the term *firmware* instead of *software,* such as "upload the firmware" or "once the firmware has been installed into the device."

Consumables

Consumables are the electronic components, the PIC microcontroller chip itself, with a few support components to get the microcontroller up and running. I recommend beginning with the 16F84 PIC microcontroller. The 16F84 is an 18-pin dip chip with 13 I/O lines and has 1K $\times$ 14 of rewritable memory. The rewritable memory allows you to reprogram the PIC chip up to 1000 times to test and troubleshoot your programs and circuits. The minimal support components are a 5-V dc power supply, oscillator (4.0-MHz crystal), and one pull-up $\frac{1}{4}$-W resistor (4.7-kΩ).

16F84 PIC Microcontroller

The PIC 16F84 microcontroller is shown in Fig. 1.4. It is a versatile microcontroller with flash memory. *Flash memory* is the terminology used to describe "rewriteable" memory. The 1K $\times$ 14-bit onboard flash memory can endure a

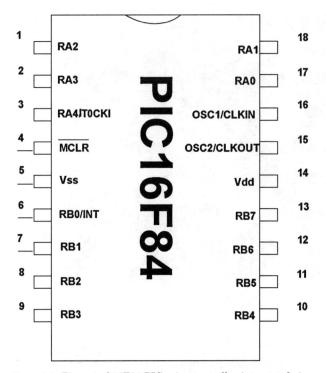

Figure 1.4 Pin-out of 16F84 PIC microcontroller integrated circuit. General features: RISC CPU 35 single-word instructions; operating speed dc, 10-MHz clock input; 1K program memory; 14-bit-wide instructions; 8-bit-wide data path; direct, indirect, and relative addressing; 1000 erase/write cycles. Peripheral features: 13 I/O pins with individual direction control; high-current sink/source for direct LED drive (25-mA sink max. per pin, 20-mA source max. per pin); TMRO—8-bit timer/counter with 8-bit programmable prescaler.

minimum of 1000 erase/write cycles. So you can reprogram and reuse the PIC chip at least 1000 times. The program retention time between erase/write cycles is approximately 40 years. The 18-pin chip devotes 13 pins to I/O. Each pin may be programmed individually for input or output. The pin status (I/O direction control) may be changed on the fly via programming. Other features include power on reset, power-saving sleep mode, power-up timer, and code protection. Additional features and architectural details of the PIC 16F84 will be given as we continue.

Step 1: Writing Code (the Basic Program)

Both the PicBasic and PicBasic Pro compilers are packaged with a free version of CodeDesigner software. CodeDesigner is an integrated development environment (IDE) for writing and programming PIC microcontrollers. CodeDesigner is an advanced text editor that is capable of calling and using both the PicBasic and PicBasic Pro compilers and the EPIC software.

If you don't want to use CodeDesigner, program text files may be written using any word processor as long as it is able to save its text file as ASCII or DOS text. If you don't own a commercial word processor, you can use Windows Notepad, which is included with Windows 3.X, 95, and 98. If you work at the DOS level, you can use the Edit program to write text files.

When you save the text file, save it with a .bas suffix. For example, if you were saving a program named wink, save it as wink.bas.

Step 2: Using the Compiler

Once set up, the CodeDesigner software will call and control the compiler and programmer software. The compiler may be run manually from a DOS window. To run the compiler program manually, enter the command pbc followed by the number of the PIC chip being programmed (that is, 16F84), then followed by the name of the source code text file. For the PicBasic Pro compiler program, the command starts with pbp instead of pbc, followed by the name of the source code text file. For example, for the PicBasic compiler, if the source code text file we created is named wink, then at the DOS command prompt enter

```
pbc -p16f84 wink.bas
```

For the PicBasic Pro compiler, the command line would be

```
pbp -p16f84 wink.bas
```

The compiler reads the text file and compiles two additional files, an .asm (assembly language) and a .hex (hexadecimal) file.

The wink.asm file is the assembly language equivalent to the Basic program. The wink.hex file is the machine code of the program written in hexadecimal numbers. It is the .hex file that is uploaded into the PIC chip.

If the compiler encounters errors when compiling the PicBasic source code, it will issue a list of errors it has found and will terminate. The errors listed need to be corrected in the source code (text file) before it will successfully compile.

Step 3: Installing the Firmware, or Programming the PIC Chip

Connect the EPIC programming board to the computer's printer port via a DB25 cable. If you are using CodeDesigner, launch the EPIC Programmer from the menu. The EPIC programming board must be connected to the parallel port and switched on before you start the software, or else the software will issue an error message "EPIC Programmer not found." Aside from the EPIC Windows software (epicwin.exe), which may be started manually in Windows or through the CodeDesigner software, there is also a DOS version of the program called epic.exe.

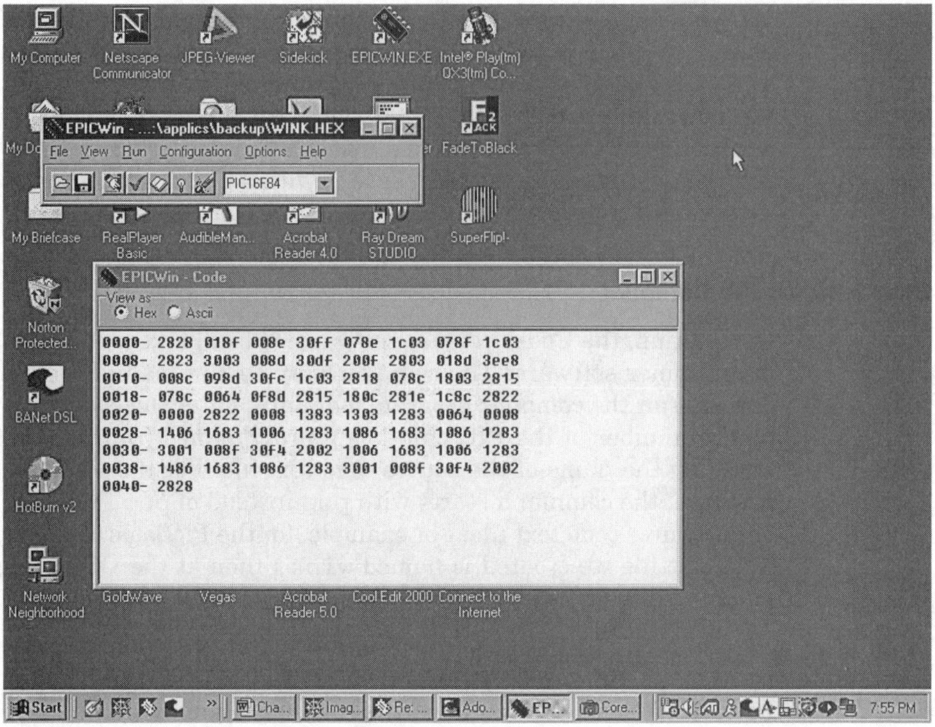

Figure 1.5   Windows version of EPIC software.

Figure 1.5 is a picture of the EPIC Windows program screen. Use the Open File option and select wink.hex from the files displayed in the dialog box. The file will load and numbers will be displayed in the code window on the left. Insert the 16F84 into the socket on the programming board, and select the Program option from the Run menu. An alternative to using the menu option is to press the Ctrl and P buttons on the keyboard. The software is then uploaded into the PIC microcontroller and is ready to be inserted into your circuit and go to work.

Ready, Steady, Go

Subsequent chapters contain step-by-step instructions for installing the software onto your hard drive and programming your first PICmicro chip.

Parts List

PicBasic Pro compiler	$249.95
PicBasic compiler	99.95
EPIC Programmer	59.95
Microcontroller (16F84)	7.95

6-ft cable (DB25)	6.95
4.0-MHz Xtal	2.50
(2) 22-pF capacitors	0.10 each

Available from Images SI Inc. (see Suppliers at end of book).
 Additional components are required in Chap. 6:

(1) Solderless breadboard	RadioShack PN# 276-175
(1) 0.1-μF capacitor	RadioShack PN# 272-1069
(8) Red LEDs	RadioShack PN# 276-208
(8) 470-Ω resistors*	RadioShack PN# 270-1115
(1) 4.7-kΩ resistor	RadioShack PN# 271-1126
(1) Voltage regulator (7805)	RadioShack PN# 276-1770
(2) Four-position PC mounted switches	RadioShack PN# 275-1301
(1) 9-V battery clip	RadioShack PN# 270-325

Available from RadioShack, Images SI Inc., Jameco Electronics, and JDR Microdevices (see Suppliers).

---

*These resistors are also available in 16-pin dip package.

2

Installing the Compiler

To compile your PicBasic programs (text files) into something that can be uploaded into the PIC microcontrollers and run, you need to first run the program text file through a compiler. So the first step is to load the compiler software onto your computer's hard drive. The following are instructions for installing the PicBasic compiler. A section on installing the PicBasic Pro compiler follows these instructions.

Installing the PicBasic Compiler Software

The first thing you need to do is to create a subdirectory on your computer's hard drive for the PicBasic compiler software. I will use Windows Explorer (Windows 95, 98, ME, 2000, XP) to create this directory. Windows Explorer can be found in the Programs folder in Windows 95 and 98 (see Fig. 2.1). For Windows ME, 2000, and XP users, Windows Explorer can be found in the Accessories folder (see Fig. 2.2).

Create a subdirectory called PBC on the computer's hard drive; then copy the files from the diskette into it. For the conventions in this book it is assumed that the reader's hard drive is drive letter C.

Start the Windows Explorer program. Highlight your computer's hard drive (usually the C drive) in the Folders window. Next highlight the File menu, then New menu, and click on the Folder option (see Fig. 2.3). Enter the name PBC in the New Folder icon (see Fig. 2.4).

Place the 3.5-in PicBasic compiler diskette into your computer's floppy drive, usually the A drive. Highlight the A drive in Windows Explorer's Folder window (see Fig. 2.5). All the files on the 3.5-in diskette will be displayed in the right-side area. Select all the files, go to Edit menu options, and choose Copy (see Fig. 2.6). Next select the PBC directory on the left side of the Windows Explorer window. Then go back to the Edit menu and select the Paste option. All the files and subdirectories on the 3.5-in diskette will be copied into the PBC directory on the hard drive.

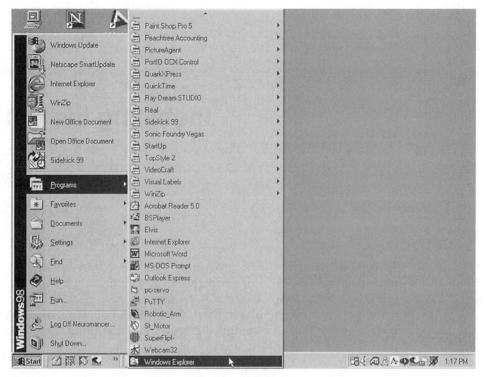

Figure 2.1   Finding Windows Explorer in Windows 95 and 98.

An alternate to pasting the selected files is to select all the files as before, copy the files, drag the selected files to the PBC directory using the mouse, and then release the mouse button (see Fig. 2.7).

Installing the PicBasic Pro Compiler

Installing the PicBasic Pro compiler is not the same procedure as outlined for the PicBasic compiler. To install the PicBasic Pro compiler, you must execute a self-extracting program that decompresses the necessary programs and files. It is recommended that you create a subdirectory named PBP on your computer's hard drive.

Start the Windows Explorer program. Highlight your computer's hard drive (usually the C drive) in the Folders window. Next highlight the File menu, then New menu, and click on the Folder option (see Fig. 2.3). Enter the name PBP in the New Folder icon (see Fig. 2.4).

Place the 3.5-in PicBasic Pro Compiler diskette into your computer's floppy drive, usually the A drive. Now here's where the installation procedure changes. For those using Windows 95 or 98, start an MS-DOS Prompt window. Click on Start, select Programs, then click on MS-DOS Prompt (see Fig. 2.8). For Windows ME, 2000, and XP users, start a Command Prompt window

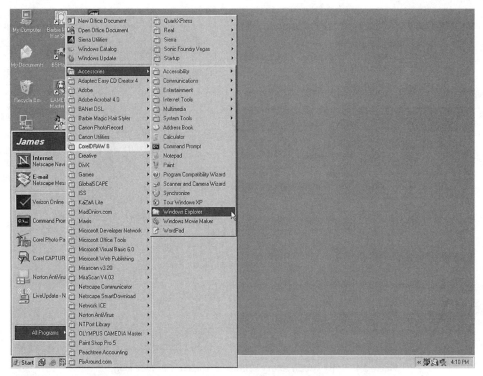

Figure 2.2 Finding Windows Explorer in Windows 2000 and XP.

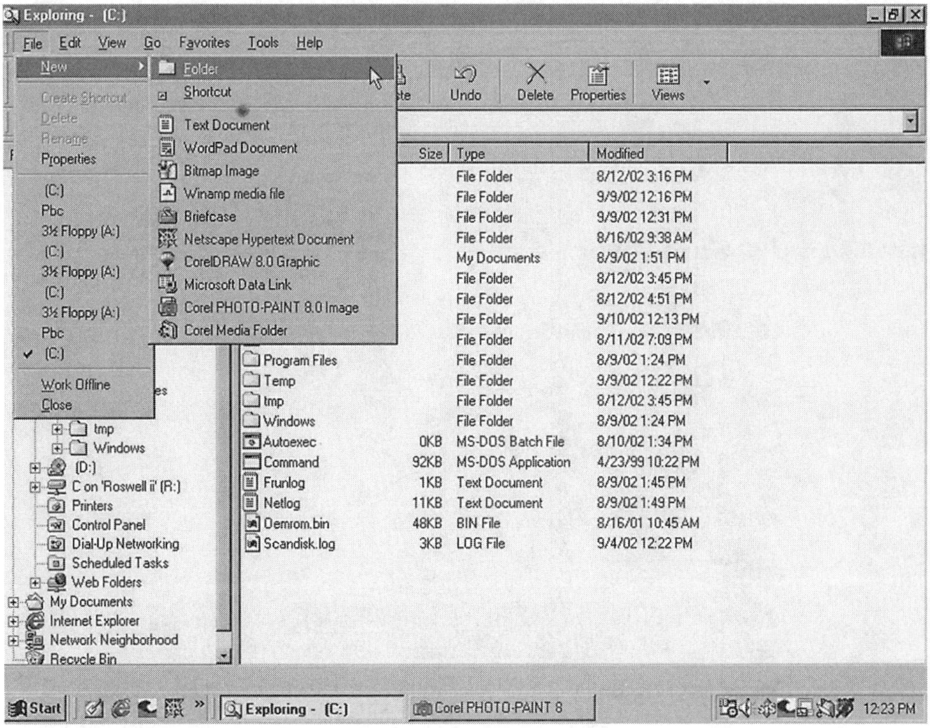

Figure 2.3 Creating a new folder (subdirectory) on computer's hard drive C.

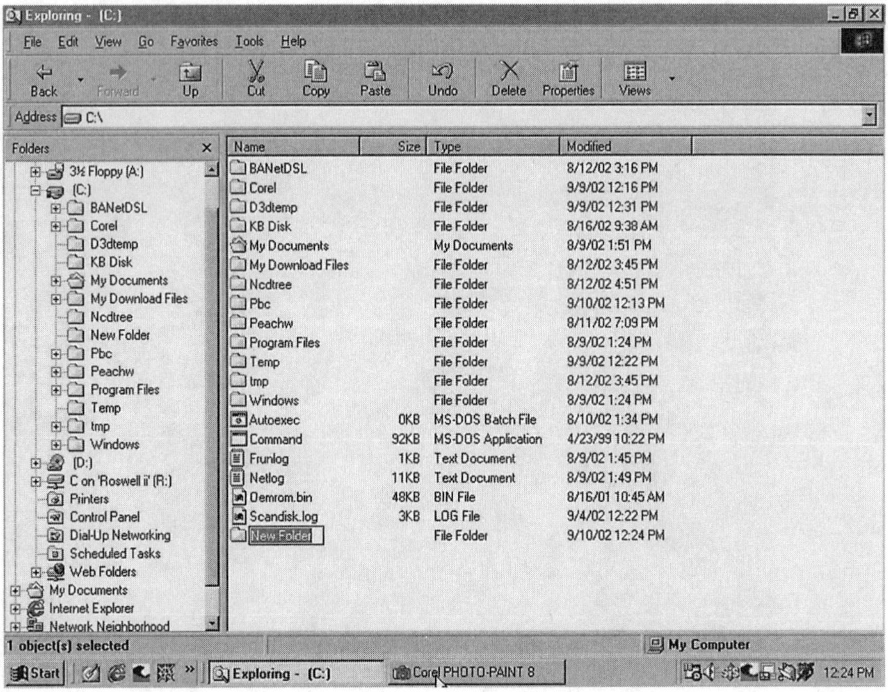

Figure 2.4   Type subdirectory's name in the New Folder icon.

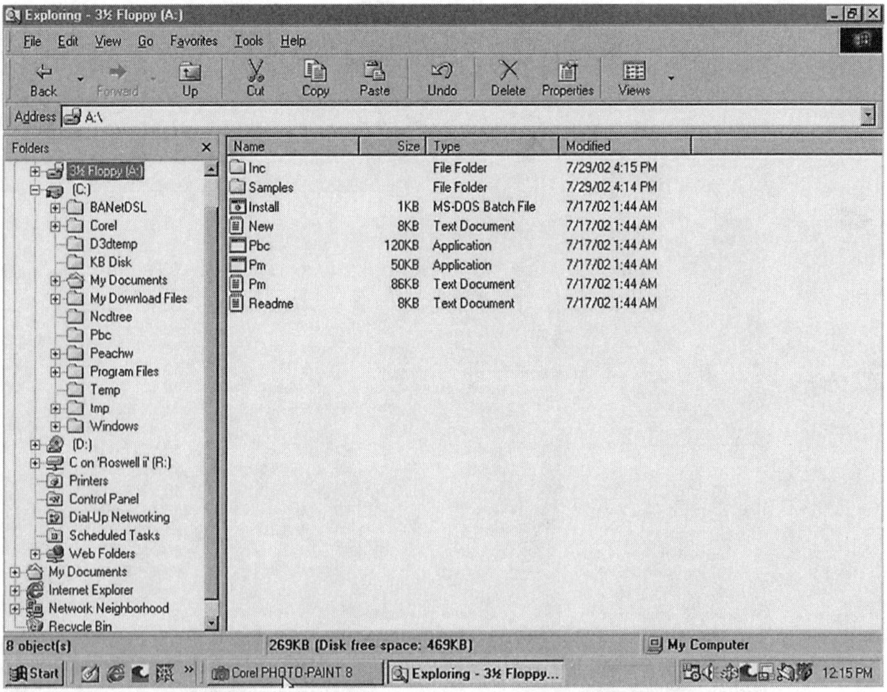

Figure 2.5   Selecting the A drive containing the PicBasic program diskette.

14

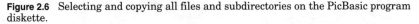

Figure 2.6 Selecting and copying all files and subdirectories on the PicBasic program diskette.

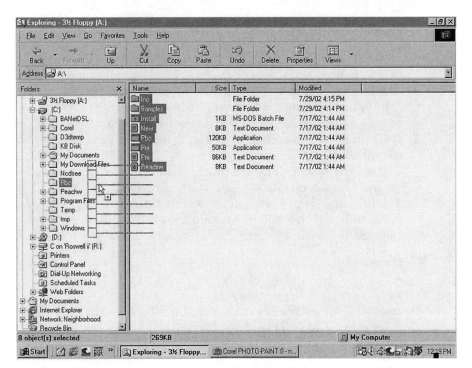

Figure 2.7 Using mouse to copy all selected files on the PicBasic program diskette in the A drive to the PBC directory on the hard drive.

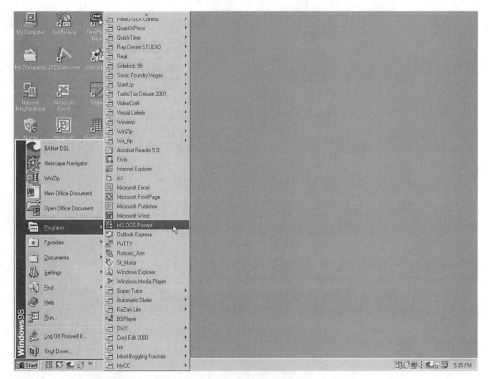

Figure 2.8   Starting MS-DOS Prompt window in Windows 95 and 98.

(equivalent to an MS-DOS Prompt window). Click on All Programs, select Accessories, and then click on Command Prompt (see Fig. 2.9).

In either the Command Prompt window or MS-DOS window, you will need to type in and use a few old-fashioned DOS commands. DOS commands are typed in on the command line, and then the Return key is hit to execute the command.

The DOS instructions are provided to help the reader and serve as a supplement to the installation directions provided with the software packages. The instructions are not meant as a DOS tutorial. More information on DOS commands can be found in any number of DOS manuals. Here is a list of DOS commands we will be using and what action they perform:

Command	**Action**
cd	Change directory
md	Make directory
copy	Copy files
xcopy	Copy files and subdirectories
path	Set a search path for executable files
dir	Directory

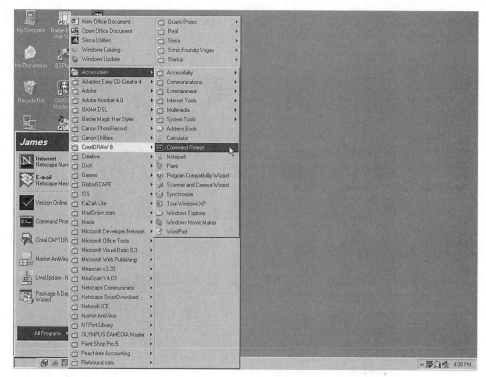

Figure 2.9 Starting Command Prompt window in Windows 2000 and XP.

From this point on, the MS-DOS Prompt window and the Command Prompt window will be referred to as the DOS window. When the DOS window is opened, you will be located in a subdirectory on the hard drive. Your prompt may look like this: C:\WINDOWS>.

The DOS prompt provides vital information. The C: tells us we are on the C drive. The \WINDOWS tells us we are in the Windows subdirectory.

We want to work from the root directory of the computer's hard drive (usually the C drive). We accomplish this by using the cd (change directory) command.

The cd.. command brings one up a single level in the directory hierarchy. Using the cd\ command brings one up to the root directory regardless of how deep (levels) one has moved into subdirectories. The root directory is the top of the directory hierarchy. From the Windows subdirectory type in cd\ and hit the Enter key to move to the root directory of the hard drive. Type in (enter) the following command and hit the Enter key.

```
cd\
```

We already created our subdirectory PBP by using Windows Explorer for the PicBasic Pro compiler. We want to move into the PBP subdirectory, enter the following command, and hit Enter.

Figure 2.10 Using DOS commands in DOS Prompt window to execute PicBasic Pro installation program.

```
c:\> cd pbp
```

Next place the 3.5-in PicBasic Pro diskette into your A drive, and type the following at the DOS prompt:

```
c:\pcp> a:\pbpxxx -d
```

Here xxx is the version number of the compiler on the disk (see Fig. 2.10). This command copies and installs all the required files into the PBP directory. With the files safely loaded onto your hard drive, remove the diskette and store it in a safe place, in case it is needed in the future.

The PicBasic Pro program is now installed. You may close the DOS window and store the 3.5-in diskette.

3

Installing the EPIC Software

Installing the EPIC software from Windows is simple. To install, run the `install.bat` file on the 3.5-in EPIC diskette. The `install.bat` file executes the main self-extracting program that automatically creates a subdirectory EPIC on your computer's hard drive, then decompresses the program and its support files, and copies them into the EPIC subdirectory.

If a subdirectory called EPIC already exists on your hard drive, when you run the `install.bat` file, you will receive an error message.

If you are still in the same DOS session as in last chapter and want to continue to use DOS to install the EPIC software, skip down to the "Installing the EPIC Software, DOS Version" section. For those who wish to use Windows to install this software, continue to read.

Installing the EPIC Software in Windows

From Windows click on the Start button, then Run (see Fig. 3.1). Place the EPIC programming diskette into the A drive. When the Run menu window opens, select Browse. From the Browse window select the A drive. This action will list the files on the A drive. Select the `install.bat` file and click the Open button (see Fig. 3.2).

This action brings you back to the Run window. The `install.bat` file should be listed in the window. See Fig. 3.3. Click on the OK button. This action automatically opens a DOS window and starts the executable program on the EPIC diskette. The executable program creates a new subdirectory on the computer's hard drive called EPIC. It decompresses and copies all the necessary files into the EPIC subdirectory, as shown in Fig. 3.4.

If you just installed the EPIC program using Windows, skip over the next section, "Installing the EPIC Software, DOS Version" and continue reading at the "Supplemental—Applications Directory" section.

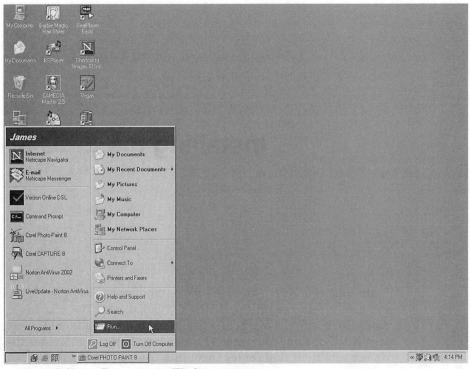

Figure 3.1 Selecting Run option on Windows start menu.

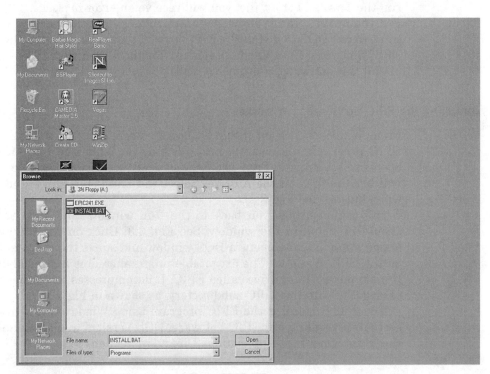

Figure 3.2 Selecting `install.bat` file on EPIC diskette.

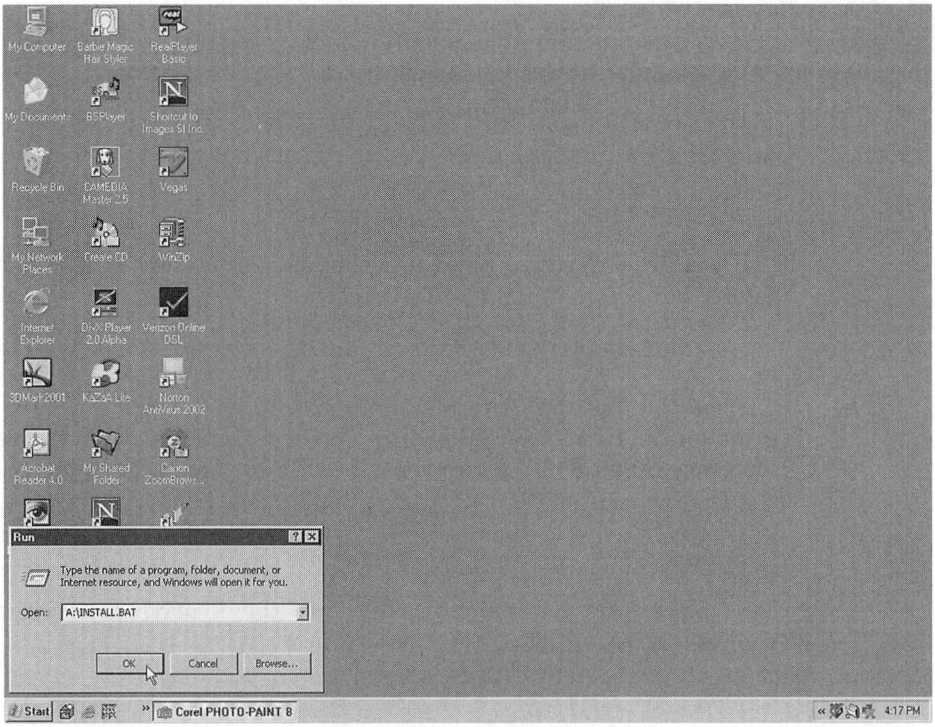

Figure 3.3 Hitting OK on `install.bat` to begin execution.

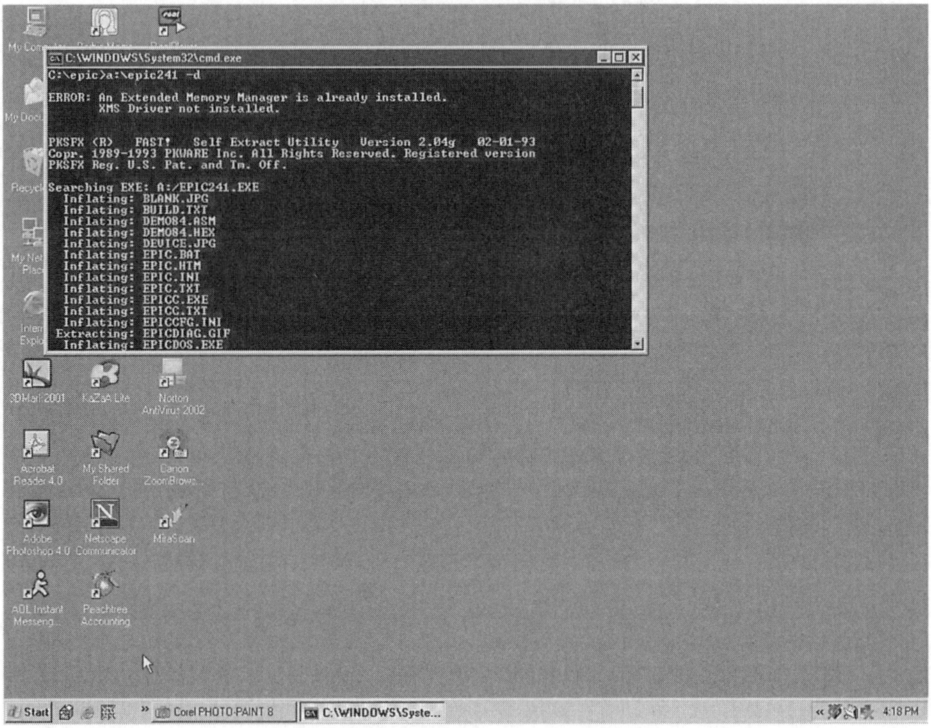

Figure 3.4 Self-extracting EPIC program running in MS-DOS window.

Installing the EPIC Software, DOS Version

If you are still operating in the same DOS session as in Chap. 2, move back into the root directory, and enter at the prompt

```
c:\> pbp cd..
```

If you are entering a new DOS window, the prompt may appear a little different, but the command is the same.

```
c:\> windows cd/
```

From the root directory of the C drive we will run the install.bat program on the EPIC 3.5-in diskette. The self-extracting file creates itself a subdirectory called EPIC. Place the 3.5-in EPIC diskette into the floppy drive. At the DOS prompt enter

```
c:\> a:
```

This places the command prompt into the A drive; the command prompt should look like this:

```
a:\>
```

Now run the install.bat file by entering the following command:

```
a:\> install.bat
```

This starts the self-extracting file that creates the EPIC subdirectory and installs all the required files into the subdirectory.

With the program and files installed onto your hard drive, remove the diskette and store it in a safe place, in case it is needed in the future.

Supplemental—Applications Directory

It would be a good idea at this time if we created another subdirectory where we can store all our PicBasic application programs. This will keep the PBC (or PBP) and EPIC directories clean, neat, and uncluttered with programs and program revisions.

From Windows Explorer create an Applics subdirectory on your computer's hard drive.

4

CodeDesigner

In this chapter we will set up and work with the CodeDesigner software. CodeDesigner is a Windows integrated development environment (IDE) interface for the PIC series of microcontrollers. This IDE interface allows one to write code, compile the code, and then program the code into a PIC microcontroller while staying in the same Windows environment.

The compiling of code within CodeDesigner still requires the use of one of the PicBasic compilers. Programming the compiled code into a PIC microcontroller requires the EPIC software and hardware. CodeDesigner integrates these software and hardware packages so that they can operate within its Windows environment.

CodeDesigner has many useful features that help you write code and that make it far superior to using a simple text editor.

CodeDesigner Features

- AutoCodeCompletion. CodeDesigner makes writing code much easier with smart pop-up list boxes that can automatically fill in statements and parameters for you.

- Multidocument support.

- Line error highlighting. Compile your PicBasic project and CodeDesigner will read error data and highlight error lines.

- QuickSyntaxHelp. The QuickSyntaxHelp feature displays statement syntax when you type in a valid PicBasic statement.

- Statement description. Statement descriptions are displayed in the status bar when you type in a valid PicBasic statement.

- Statement Help. Simply position your cursor over a PicBasic statement and get statement-specific help.

- Label listbox. The label listbox displays the current label and allows you to select a label from the list to jump to.

- Colored PicBasic syntax. This sets colors for reserved words, strings, numbers, comments, defines, etc. Colored PicBasic syntax makes for easy code reading.

- Bookmarks. Never lose your place again. CodeDesigner allows you to set bookmarks.

- Multiple undo/redo actions. If you didn't want to delete that last line, it's no problem. Simply click on the Undo button.

- Multiple views. Multiple views of your source code allow you to easily edit your code.

- Print source code.

- Drag-and-drop text.

- Row/column-based Insert, Delete, and Copy.

- Search and replace.

- Compile and launch device programmer.

One feature I like is that each typed line of code is color-coded, making it easier to spot errors and read through your code.

When you purchase either the PicBasic or PicBasic Pro compilers, it is packaged with an additional diskette that contains a free version of CodeDesigner called CodeDesigner Lite. The Lite version allows you to write programs up to 150 lines and open up three source files at once for easy copy and paste. If you would like to try CodeDesigner without purchasing a compiler, CodeDesigner Lite is freely downloadable from the Internet (see Parts List at end of chapter).

The idea is, if you like the free CodeDesigner software, you can then upgrade to the full-featured CodeDesigner. The standard version of CodeDesigner costs $75.00 and removes the restrictions imposed in the Lite version. This standard version allows you to write programs with an unlimited amount of code lines and to open an unlimited amount of source files. Of course, *unlimited* means with respect to the limits of your computer's capabilities.

If for any reason someone does not wish to use the CodeDesigner software, the procedures for writing code and compiling and programming a PICmicro chip manually from a DOS environment are covered in Chap. 5.

CodeDesigner increases productivity and the ease with which you can write, debug, and load PicBasic programs into the microcontroller. If there is a problem (more often than not), debugging the code and recompiling are much easier and faster when you are using CodeDesigner. When the program is completely debugged, it can be uploaded into the PIC microcontroller via the EPIC software and programming board. At this point the microcontroller and circuit are tested. If they function properly, I'm finished; if not, I begin rewriting the program or redesigning the electronics.

Software Installation

The CodeDesigner software loads as most standard Windows software does. Load the software on your computer's hard drive according to the instructions given with the software.

When CodeDesigner installs, it creates a subdirectory in the Program Files directory. It places a CodeDesigner shortcut on the Start, Program menu in Windows.

Setting CodeDesigner Options

In order for CodeDesigner to compile code and program the resulting code into PIC microcontrollers, we need to configure the default locations where CodeDesigner looks for its support programs. We set up the default locations by entering the software paths where CodeDesigner stores programs, looks for the PicBasic compiler and where to find the EPIC program.

Start the CodeDesigner software (see Fig. 4.1); the evaluation copy opens with this version of the window. The next window is the standard opening screen to the CodeDesigner software (see Fig. 4.2). To begin setting the options, click on the Compile menu option and then on Compiler Options (see Fig. 4.3).

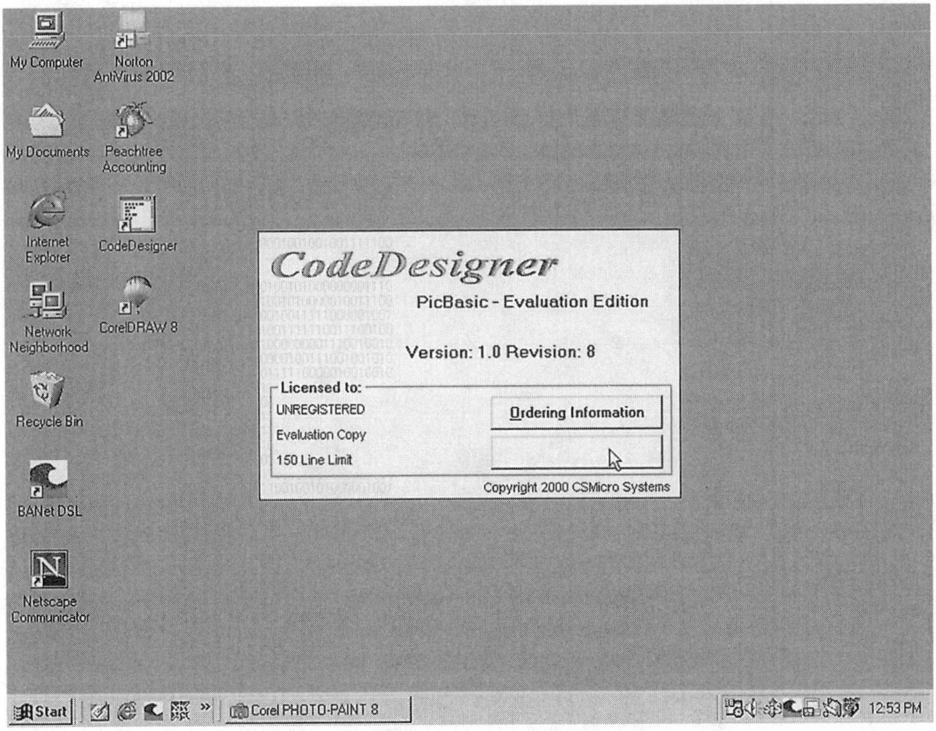

Figure 4.1 CodeDesigner Lite version start-up screen.

Figure 4.2 Program written in CodeDesigner ready for compiling and programming.

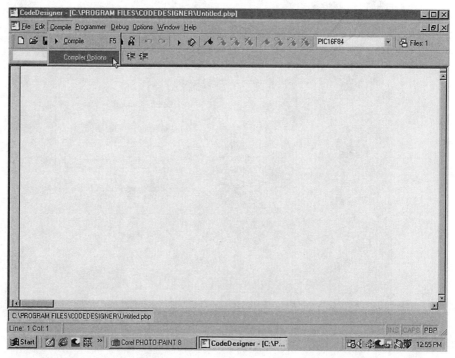

Figure 4.3 Error created when Windows cannot identify EPIC Programmer attached to printer port.

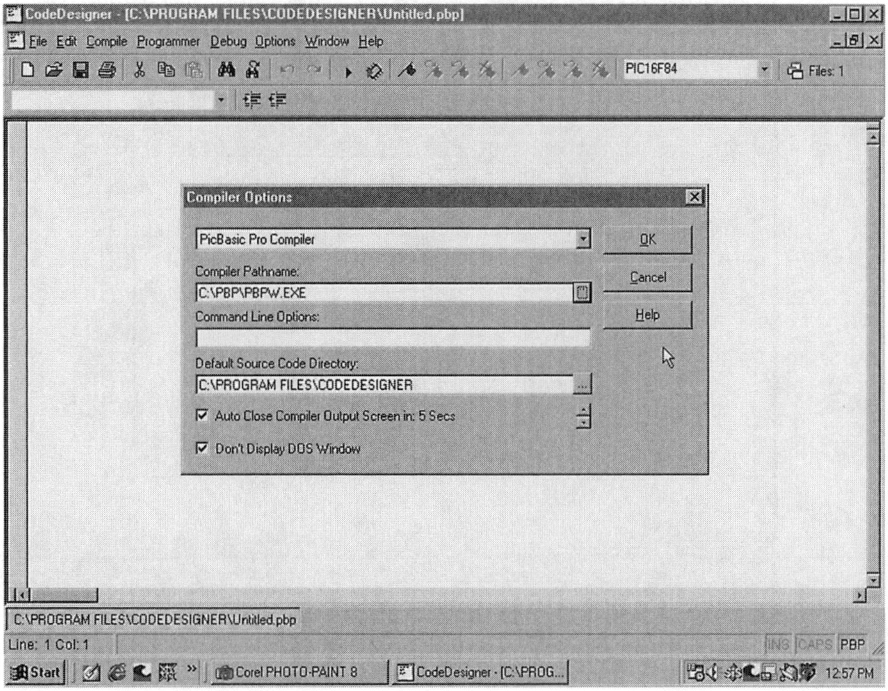

Figure 4.4 Schematic for wink program.

The Compiler Options window opens (see Fig. 4.4). In the top text field, use the pull-down menu to select which compiler you are using, the PicBasic Pro or PicBasic. In Fig. 4.4 the PicBasic Pro compiler is chosen.

In the second text field, you select the compiler's pathname. The compiler path and name (pbpw.exe) is chosen for the PicBasic Pro compiler, in the subdirectory of C:\PBP.

In the third text field we choose where the CodeDesigner software looks to load and save our source code files. Hit the Browse button next to the text field. This opens a browser window (see Fig. 4.5); select the Applics subdirectory on the hard drive and click on OK.

The "Default Source Code Directory" text field now contains the path "C:\Applics" subdirectory (see Fig. 4.6). Click the OK button to close the Compiler Options window .

Now we need to set the Programmer Options. Click on the Programmer, Programmer Options on the top menu (see Fig. 4.7). This opens the Programmer Options window (see Fig. 4.8). Click on the Browse button next to the Programmer Pathname text field. A browser window opens; select the epicwin.exe program in the EPIC subdirectory on your computer's hard drive (see Fig. 4.9). Click Open, and this brings you back to the Programmer Options window. The new path you select should be in the Programmer Pathname text field (see Fig. 4.10). Click OK to set this option.

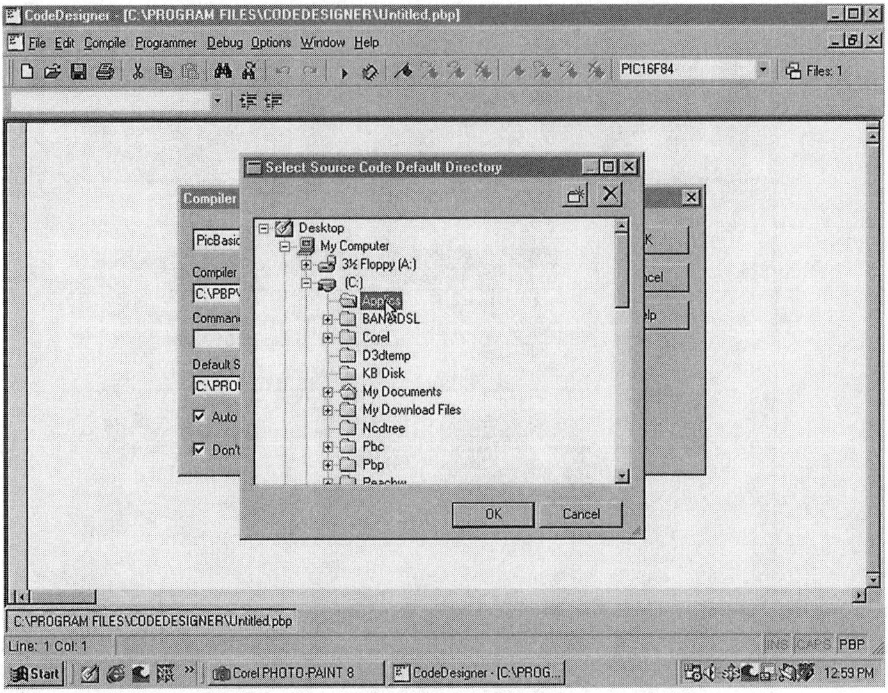

Figure 4.5 Complete circuit built on solderless breadboard.

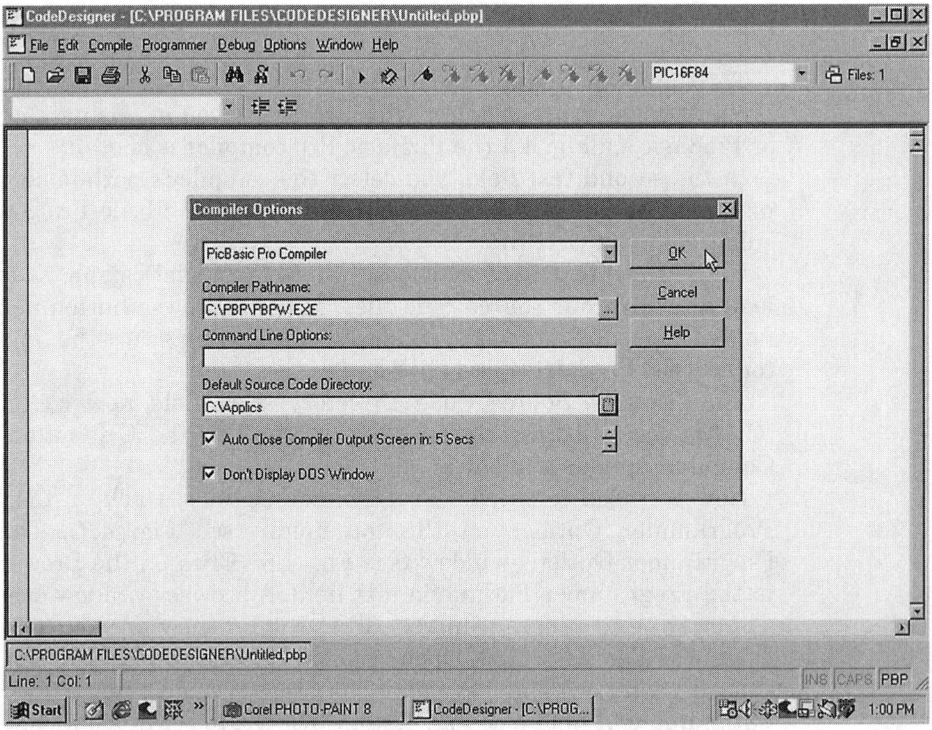

Figure 4.6 Selecting compiler (CodeDesigner).

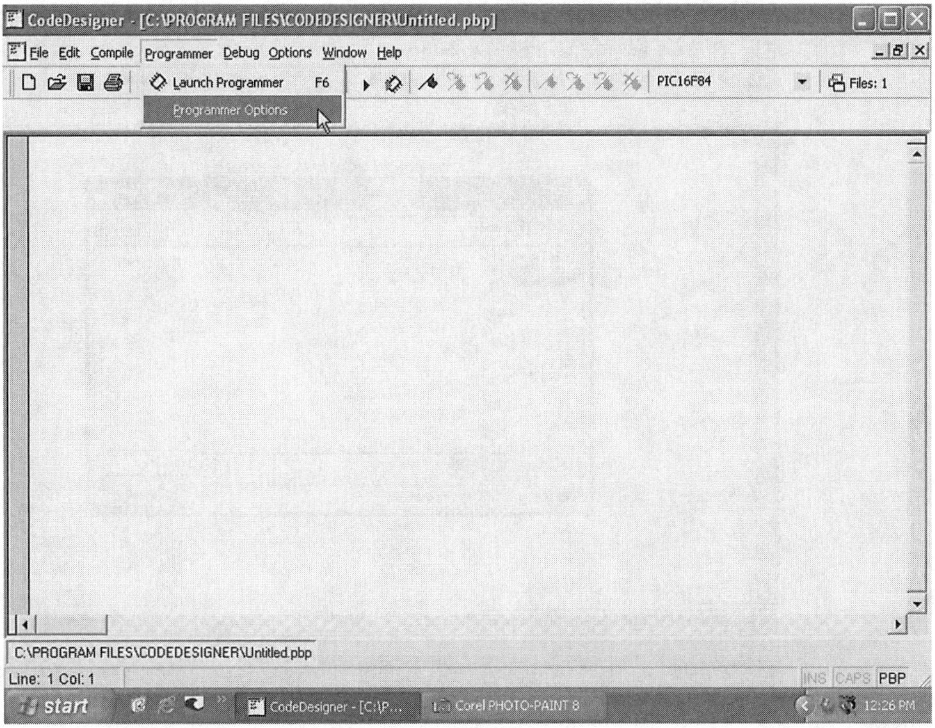

Figure 4.7 Opening Programmer options (CodeDesigner).

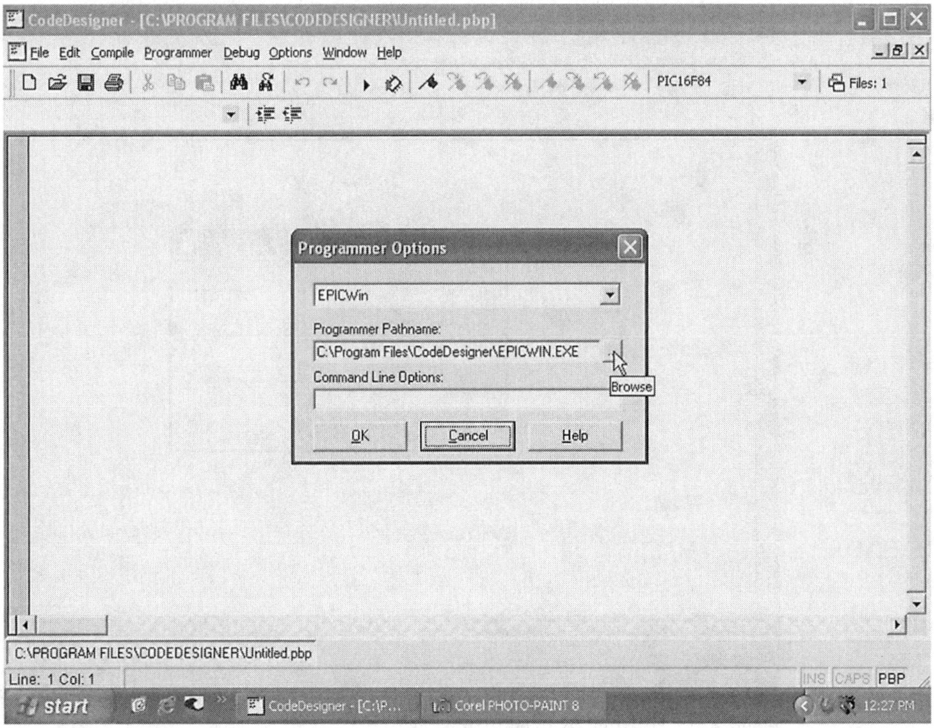

Figure 4.8 Programmer options window (CodeDesigner).

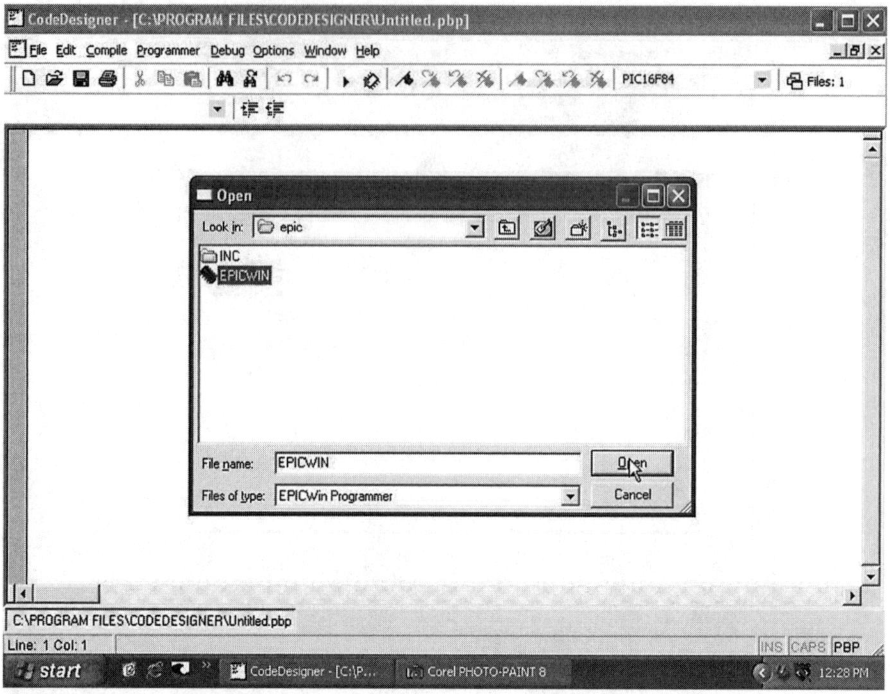

Figure 4.9 Selecting `epicwin.exe` Programmer in EPIC subdirectory.

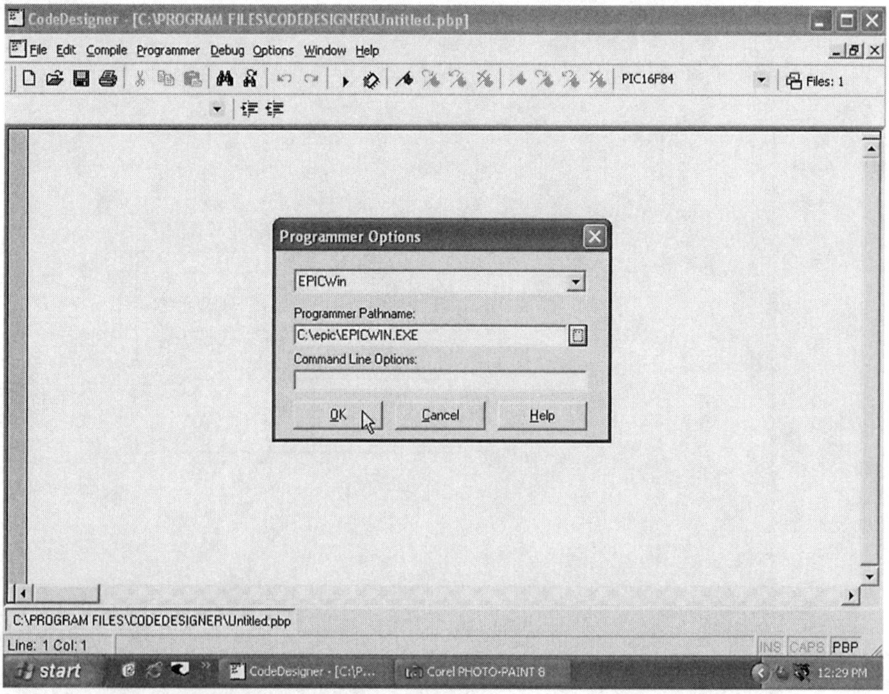

Figure 4.10 Hitting OK to confirm selection.

With the CodeDesigner options set, we are ready to write our first program.

First Program

Start CodeDesigner and enter the following code for the PicBasic compiler.

```
'First PicBasic program to wink two LEDs connected to port b.
 loop: high 0              'Turn on LED connected to pin rb0
    low 1                  'Turn off LED connected to pin rb1
    pause 500              'Delay for .5 second
    low 0                  'Turn off LED connected to pin rb0
    high 1                 'Turn on LED connected to pin rb1
    pause 500              'Delay for .5 second
    goto loop              'Go back to loop and blink & wink LEDs forever
    end
```

The next program is identical in function (not code) to the PicBasic program above. Start CodeDesigner and enter the following code (see Fig. 4.11):

```
'Wink program
'Blinks and winks two LEDs connected to port b
```

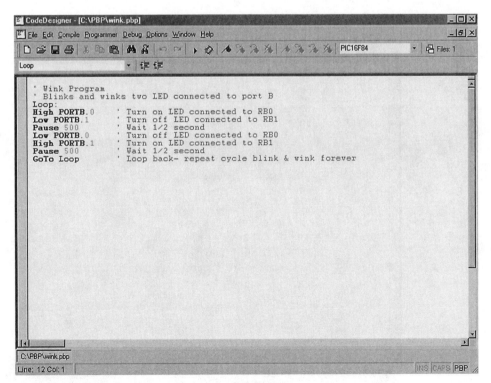

Figure 4.11 PicBasic Pro program written in CodeDesigner.

```
loop:
high portb.0                  'Turn on LED connected to rb0
low portb.1                   'Turn off LED connected to rb1
pause 500                     'Wait 1/2 second
low portb.0                   'Turn off LED connected to rb0
high portb.1                  'Turn on LED connected to rb1
pause 500                     'Wait 1/2 second
goto loop                     'Loop back--repeat cycle blink & wink forever
```

CodeDesigner defaults to writing code for the PIC 16F84 microcontroller. This is the microcontroller I recommend to start with. To change the microcontroller, simply pull down the device menu and select the appropriate microcontroller (see Fig. 4.12).

When CodeDesigner attempts to compile a program from the Windows environment, it automatically opens a DOS Prompt window, compiles the program, and then ends the DOS session.

To compile the program using CodeDesigner, either select compile under the Compile menu or hit F5. CodeDesigner automatically starts the PicBasic Pro compiler (or PicBasic compiler) to compile the program. Before you attempt to compile a program, make sure you have set up Compiler Options under the Compile menu.

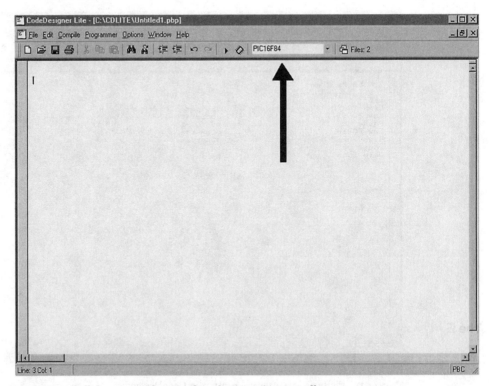

Figure 4.12 Pull-down menu location for selecting microcontroller.

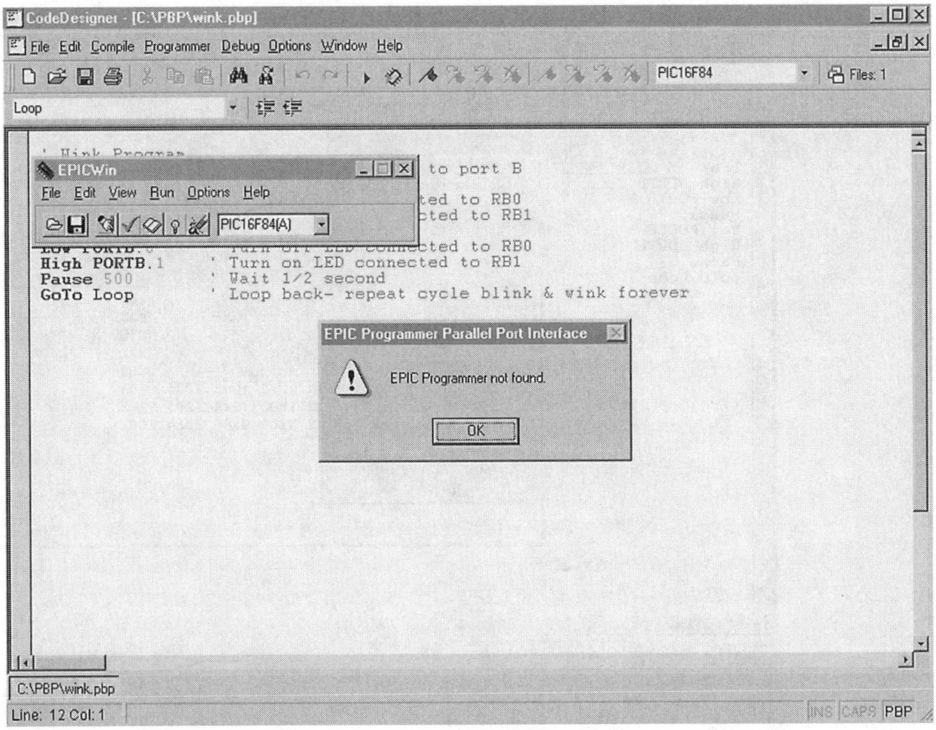

Figure 4.13 Error message generated when CodeDesigner cannot find Programmer.

Once the program is compiled, we can go to the next step of loading the program into a PIC microcontroller chip.

Connect the EPIC programming board to the printer port. If your computer has only one printer port, disconnect the printer, if one is connected, and attach the EPIC programming board, using a 6-ft DB25 cable.

When you connect the EPIC programming board to the computer, there should not be any microcontroller installed onto the board. If you have an ac adapter for the EPIC programmer board, plug it into the board. If not, attach two fresh 9-V batteries. Connect the "battery on" jumper to apply power. The programming board must be connected to the printer port with power applied to the programming board before the EPIC programming software is started. If it is not, the software will not see the programming board connected to the printer port and will give the error message "EPIC Programmer Not Found" (see Fig. 4.13).

The EPIC Programming Board Software

To program the 16F84 microcontroller from within CodeDesigner, select the Launch Programmer menu item from the Programmer menu, or hit F6. CodeDesigner automatically starts the epicwin.exe Windows software.

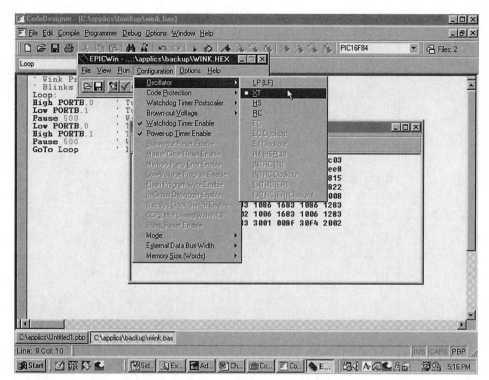

Figure 4.14 Setting configuration switches in `epicwin.exe` program.

With the EPIC Windows software started, set the configuration switches one by one under the Options menu item (see Fig. 4.14).

Device: Sets the device type. Set it for 16F84 (default).

Memory Size (K): Sets memory size. Choose 1.

OSC: Sets oscillator type. Choose XT for crystal.

Watchdog Timer: Choose On.

Code Protect: Choose Off.

Power Up Timer Enable: Choose On.

After the configuration switches are set, insert the PIC 16F84 microcontroller into the open socket on the EPIC programming board. If you received an error message "EPIC Programmer Not Found" when CodeDesigner started the EPIC Windows program (see Fig. 4.13), you have the option of either troubleshooting the problem or using the EPIC DOS program. For instructions on using the EPIC software, DOS version, see Chap. 5.

The schematic of the circuit needed to test the PICmicro is given in Chap. 6. If you have successfully written, compiled, and uploaded the code into the PICmicro chip using CodeDesigner, then you can skip the DOS material in Chap. 5 and pick up at "Testing the PIC Microcontroller" in Chap. 6.

Parts List

CodeDesigner Lite   Free

Download from Internet at: www.imagesco.com/catalog/pic/codedesigner.html.

CodeDesigner Standard $75.00

Available from Images SI Inc. (see Suppliers at end of book).

Chapter 5

Using DOS to Code, Compile, and Program

In Chap. 4 we compiled and programmed our microcontroller, using the CodeDesigner program. If for some reason you do not wish to use or cannot use CodeDesigner Lite, this chapter will instruct you in how to perform all the functions for writing code, compiling the code, and programming the code in a PICmicro chip from DOS or a DOS Prompt window.

When you start a new DOS session, use the path command (see Fig. 5.1), so that you will not have to copy and swap files back and forth across directories. If you have created the directory names as suggested in this book, you can use the following command.

For PicBasic users the command is

```
path \;c:\pbc;c:\epic;c:\windows\command;
```

For PicBasic Pro users, the command is

```
path \;c:\pbp;c:\epic;c:\windows\command;
```

Now we can begin by using a standard word processor or text editor to write the PicBasic program in DOS. Windows users can use the Notepad program. DOS-level users can use the Edit program. In DOS we will work from and store our program(s) in the subdirectory we created earlier, called Applics.

Move into the Applics subdirectory. Use the cd (change directory) command. Enter this at the DOS prompt (see Fig. 5.1):

```
c:\> cd applics
```

Once in this directory, the prompt changes to

```
c:\applics>
```

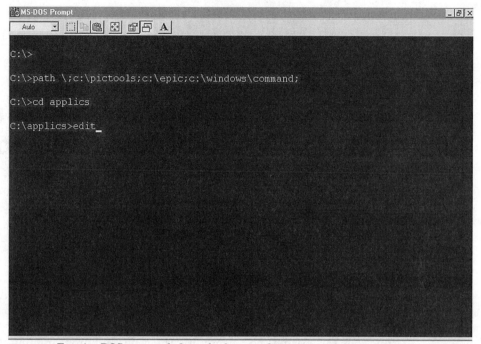

Figure 5.1 Entering DOS commands for path, changing directories, and starting the Edit program.

In this example I will be using the free Edit program package with Windows to write the program. Start Edit by typing Edit at the command prompt (see Fig. 5.1).

```
c:\applics> edit
```

This starts the Edit program (see Fig. 5.2). Enter this program in your word processor exactly as it is written:

```
'1st PicBasic program
'Wink two LEDs connected to port b.
loop: high 0              'Turn on LED connected to pin rb0
   low 1                  'Turn off LED connected to pin rb1
   pause 500              'Delay for .5 second
   low 0                  'Turn off LED connected to pin rb0
   high 1                 'Turn on LED connected to pin rb1
   pause 500              'Delay for .5 second
   goto loop              'Go back to loop and blink & wink LEDs forever
   end
```

See Fig. 5.3. Save the text file in the Applics directory. Use the Save function under the File menu. Name the file wink.bas (see Fig. 5.4). If by accident you saved the file as wink.txt, don't get discouraged. You can do a

Figure 5.2 Start screen of Edit program.

Save As from the Edit program (under File menu) and rename the file wink.bas.

For PicBasic Pro users, enter the following text in your word processor and save the file as wink.bas.

```
'1st PicBasic Pro program
'Winks two LEDs connected to port b
loop:
high portb.0            'Turn on LED connected to rb0
low portb.1             'Turn off LED connected to rb1
pause 500               'Wait 1/2 second
low portb.0             'Turn off LED connected to rb0
high portb.1            'Turn on LED connected to rb1
pause 500               'Wait 1/2 second
goto loop               'Loop back--repeat cycle blink & wink forever
```

Compile

The PicBasic compiler (or PicBasic Pro compiler) may be run from DOS or from a DOS Prompt window within Windows.

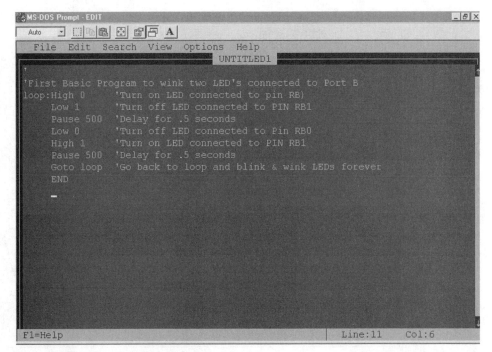

Figure 5.3 Entering wink.bas program.

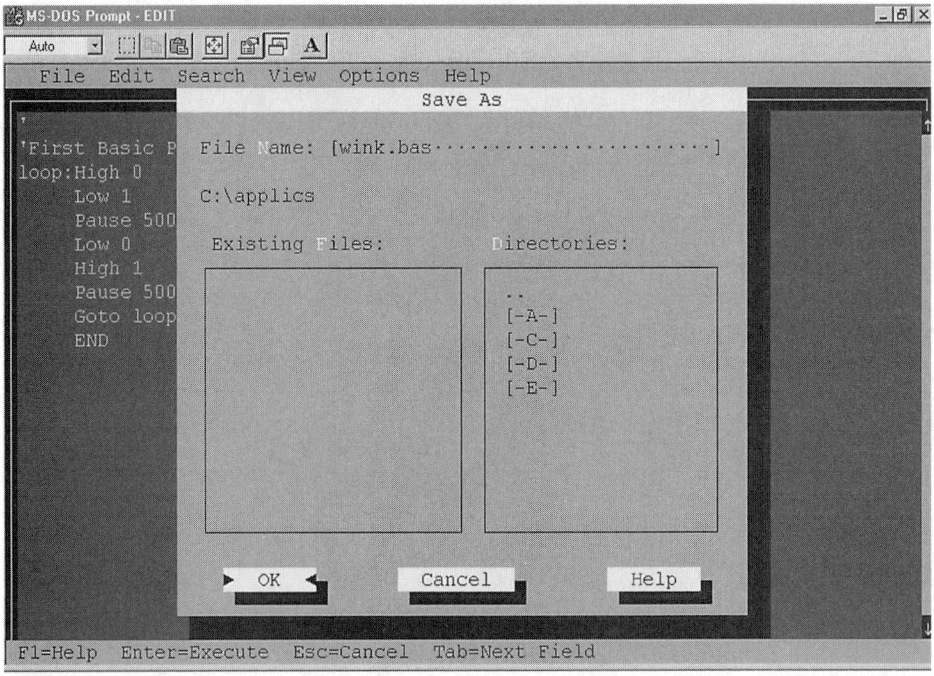

Figure 5.4 Saving wink.bas program.

```
MS-DOS Prompt                                                    _ 🗗 ×
 Auto    ▾  ▫ ▫ ▩ 🖾 🗗🗗 A

C:\>md applics

C:\>path \;c:\pictools;c:\epic;c:\windows\command;

C:\>cd applics

C:\applics>edit

C:\applics>pbc -p16f84 wink.bas_
```

Figure 5.5 Entering DOS command to run PicBasic compiler program on the wink.bas program for the 16F84 microcontroller.

We will run the PicBasic compiler from the Applics directory, type the command pbc -p16f84 wink.bas at the DOS prompt, and hit the Enter key (see Fig. 5.5).

```
c:\applics> pbc -p16f84 wink.bas
```

(For PicBasic Pro the command is c:/applics>pbp -p16f84 wink.bas.) The compiler displays an initialization copyright message and begins processing the Basic source code (see Fig. 5.6). If the Basic source code is without errors (and why shouldn't it be?), it will create two additional files. If the compiler finds any errors, a list of errors with their line numbers will be displayed. Use the line numbers in the error message to locate the line number(s) in the .bas text file where the error(s) occurred. The errors need to be corrected before the compiler can compile the source code correctly. The most common errors are with Basic language syntax and usage.

You can look at the files by using the dir directory command. Type dir at the command prompt

```
c:\applics> dir
```

and hit Enter (see Fig. 5.7).

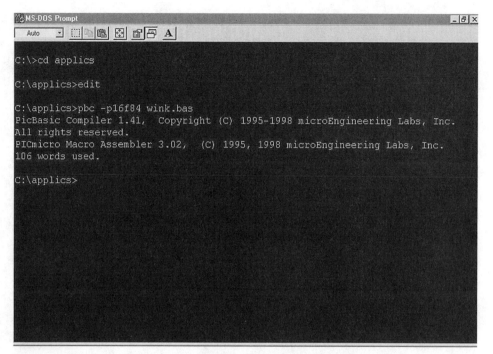

Figure 5.6 Typical copyright notice and notice provided by the PicBasic compiler when it is run successfully.

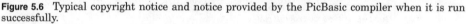

Figure 5.7 Executing DOS dir (directory) command to see the two additional files (.asm and .hex) created by the PicBasic compiler.

The `dir` command displays all the files and subdirectories within the subdirectory where it is issued. In Fig. 5.7 we can see the two additional files that the compiler created. One file is the `wink.asm` file and is the assembler source code file that automatically initiated the macroassembler to compile the assembly code to machine language hexadecimal code. The hex code file is the second file created, called `wink.hex`.

Programming the PIC Chip

To program the PIC chip, we must connect the EPIC programming carrier board (see Fig. 5.8), to the computer. The EPIC board connects to the printer port. The printer port is also called the parallel port. Either name may be used; they are both correct. A computer may contain up to four parallel (printer) ports. Each port is assigned a number, from 1 through 4. The computer lists these ports as LPT1 to LPT4.

If your computer has only one printer port, disconnect the printer, if one is connected, and attach the EPIC programming board using a 6-ft DB25 cable. In some cases it may be necessary to temporarily remove the printer driver. Figure 5.9 shows a typical window to disable an HP printer.

When you are connecting the programming board to the computer, make sure there are no PIC microcontrollers installed onto the board. If you have an ac adapter for the EPIC Programmer board, plug it into the board. If not, attach two fresh 9-V batteries. Connect the "battery on" jumper to apply pow-

Figure 5.8 EPIC programming board and software.

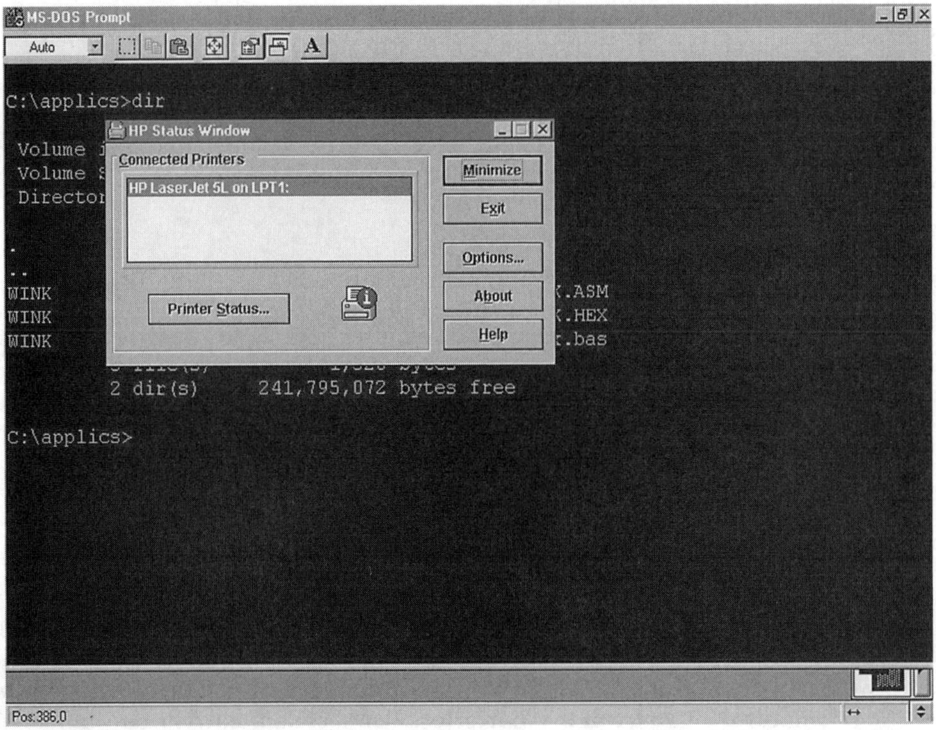

Figure 5.9 Message box window used to temporarily close printer driver to provide better access to printer port for EPIC programmer.

er. The programming board must be connected to the printer port with power applied to the programming board before the software is started. If not, the software will not see the programming board connected to the printer port and will give the error message "EPIC Programmer Not Found."

When power is applied and it is connected to the printer port, the LED on the EPIC Programmer board may be on or off at this point. Do not insert a PIC microcontroller into the programming board socket until the EPIC programming software is running.

The EPIC Programming Board Software

There are two versions of the EPIC software: `epic.exe` for DOS and `epicwin.exe` for Windows. The Windows software is 32-bit. It may be used with Windows 95, 98, NT, and XP, *but not* 3.X. It has been my experience that Windows 95 and 98 printer drivers many times like to retain control of the printer (LPT1) port. If this is the case with your computer, the Windows EPIC program may not function properly, and you may be forced to use the DOS-level program. If you receive an error message "EPIC Programmer Not Found" when you start the EPIC Windows program, you have the option of either troubleshooting the problem or using the EPIC DOS program.

Using EPIC DOS version

If using Windows 95 or higher, you could either open a MS-DOS Prompt window or restart the computer in the DOS mode. Windows 3.XX users end the Windows session.

Continuing with the `wink.bas` program

Assume we are still in the same DOS session and have just run the PBC compiler on the `wink.bas` program. We are still in the Applics directory. At the DOS prompt, type EPIC and hit Enter to run the DOS version of the EPIC software (see Fig. 5.10).

If you are operating out of a DOS window, you may get a device conflict message box, shown in Fig. 5.11. We want MS-DOS to control the LPT port so the EPIC programming software will work. Select the MS-DOS Prompt and hit the OK button.

EPIC's opening screen is shown in Fig. 5.12. Use the mouse to click on the Open button, or press ALT-O on your keyboard. Select the `wink.hex` file (see Fig. 5.13). When the `.hex` file loads, you will see a list of numbers in the win-

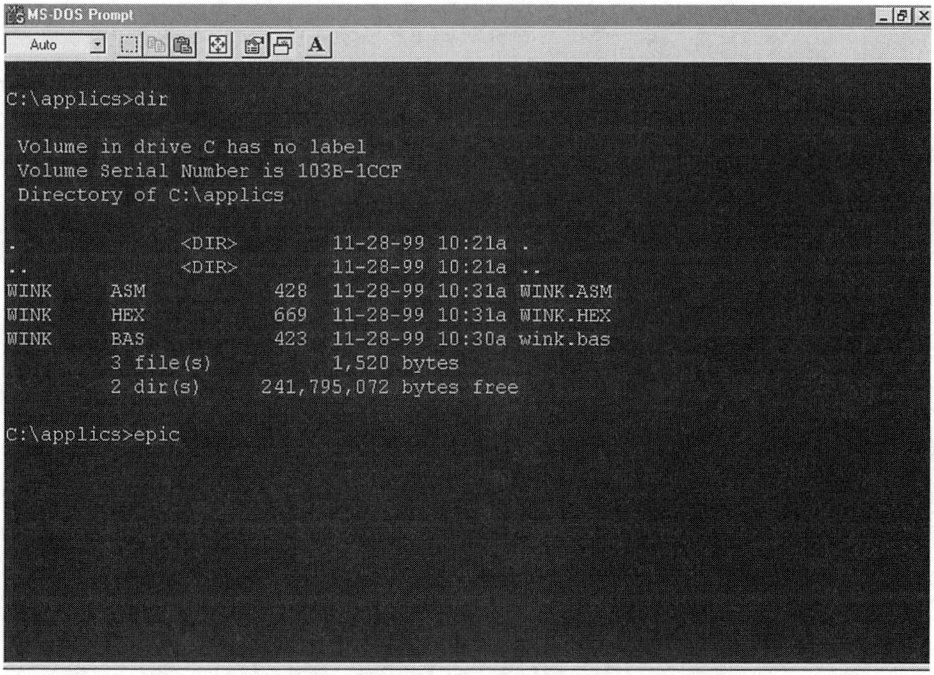

```
MS-DOS Prompt                                                    _ |8|X
 Auto      :: [] [][] [] []
C:\applics>dir

 Volume in drive C has no label
 Volume Serial Number is 103B-1CCF
 Directory of C:\applics

.              <DIR>         11-28-99 10:21a .
..             <DIR>         11-28-99 10:21a ..
WINK     ASM           428   11-28-99 10:31a WINK.ASM
WINK     HEX           669   11-28-99 10:31a WINK.HEX
WINK     BAS           423   11-28-99 10:30a wink.bas
         3 file(s)          1,520 bytes
         2 dir(s)     241,795,072 bytes free

C:\applics>epic
```

Figure 5.10   Entering DOS EPIC command to start program.

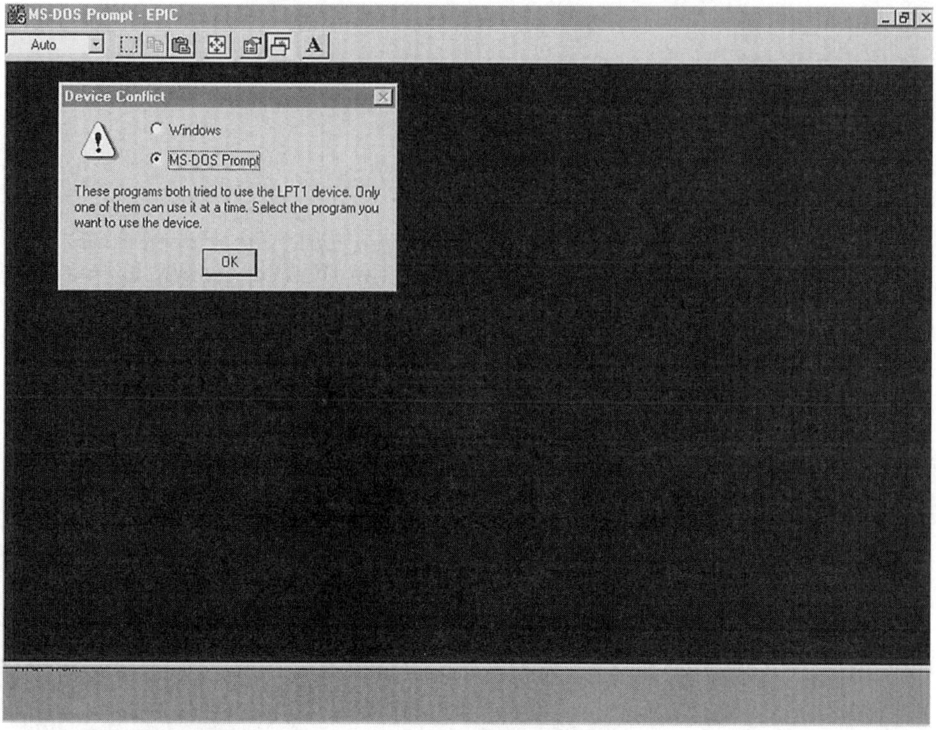

Figure 5.11 Possible "Device Conflict" error message when DOS and Windows both try to use printer port. Select DOS and hit OK button.

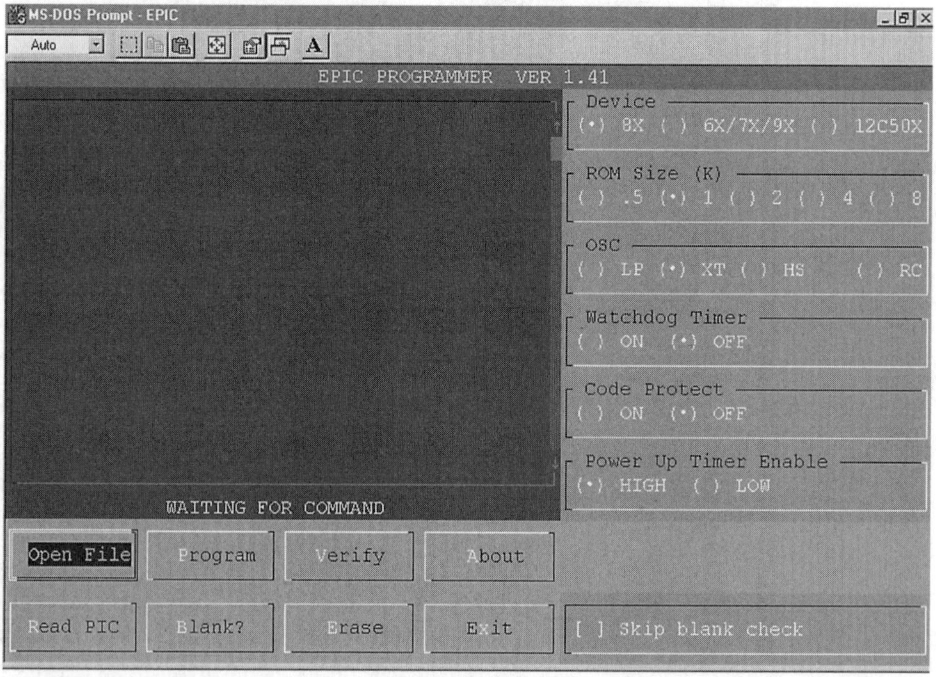

Figure 5.12 Opening screen of the EPIC programming software. Hit Open File button.

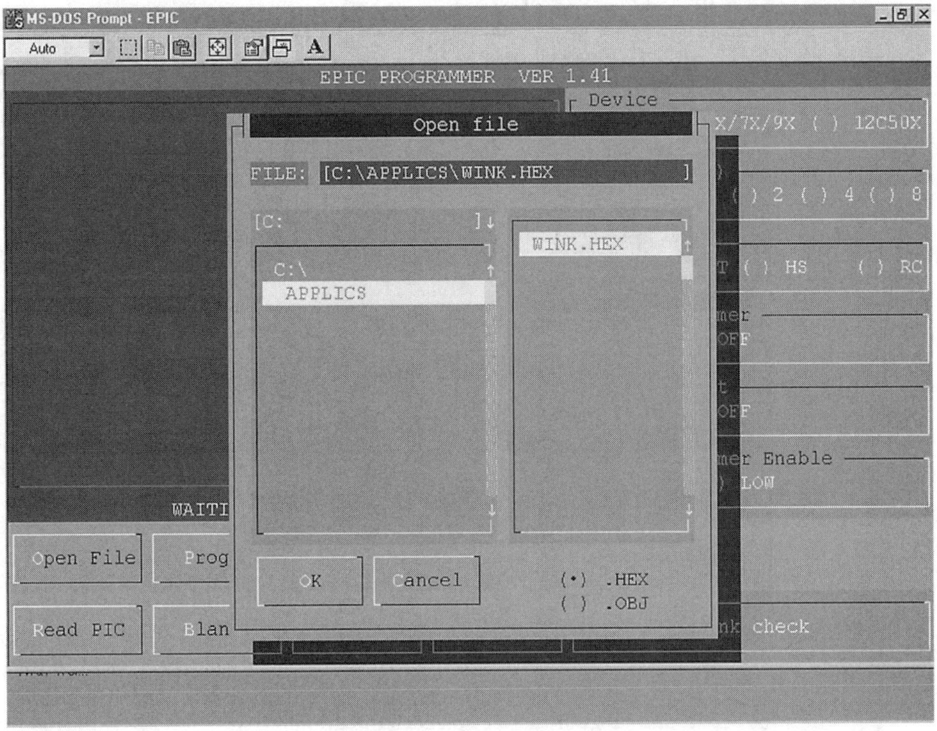

Figure 5.13   Select wink.hex in Open File message box and hit the OK button.

dow on the left (see Fig. 5.14). This is the machine code of your program. On the right-hand side of the screen are configuration switches that we need to check before we program the PIC chip.

Let's go through the configuration switches once more.

Device: Sets the device type. Set it for 8X.

Memory Size (K): Sets memory size. Choose 1.

OSC: Sets oscillator type. Choose XT for crystal.

Watchdog Timer: Choose On.

Code Protect: Choose Off.

Power Up Timer Enable: Choose On.

After the configuration switches are set, insert the PIC 16F84 microcontroller into the socket. Click on Program or press ALT-P on the keyboard to begin programming. The EPIC program first looks at the microcontroller chip to see if it is blank. If the chip is blank, the EPIC program installs your program into the microcontroller. If the microcontroller is not blank, you are given the options to cancel the operation or overwrite the existing program with the new program. If there is an existing program in the PIC chip's memory, write over it.

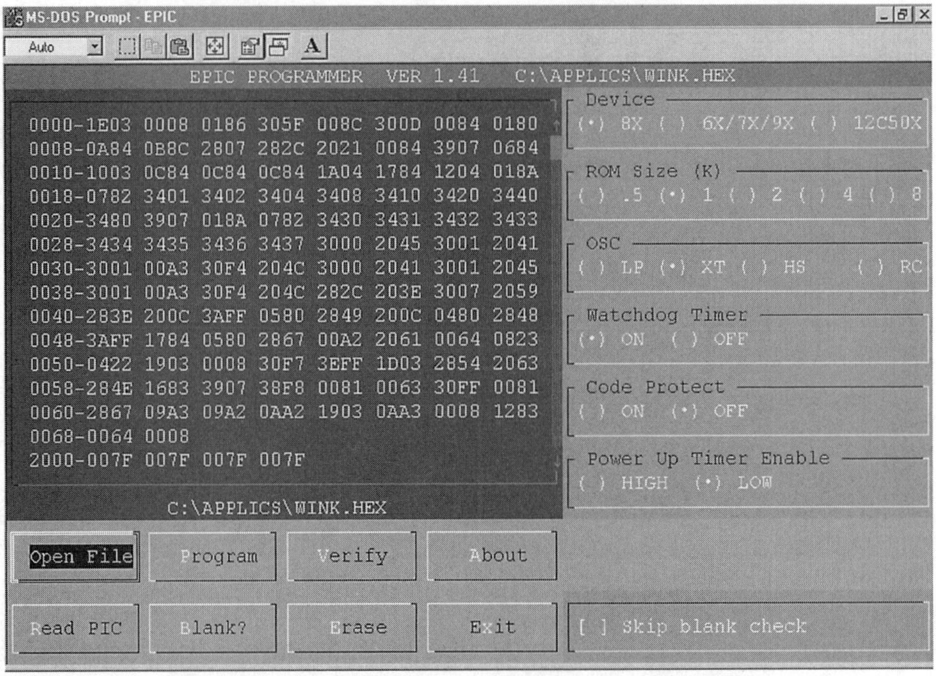

Figure 5.14 Hexadecimal numbers showing in EPIC window are the machine language version of the wink.bas program that is uploaded (programmed) into the 16F84 microcontroller.

I have noticed that when I place a brand new PICmicro 16F84 chip into the EPIC compiler to program, EPIC always reports existing code on the chip. I don't know if Microchip Technology Inc. loads numbers into the chip's memory for testing purposes. Don't let it throw you—the PICmicro chip is new.

The machine language code lines are highlighted as the EPIC software uploads the program into the PICmicro chip. When it is finished, the microcontroller is programmed and ready to run. You can verify the program if you like by hitting (or highlighting) the Verify button. This initiates a comparison of the program held in memory to the program stored in the PIC microcontroller.

Testing the PIC Microcontroller

The PIC Microcontroller

This is where we will build the testing circuit for the PICmicro chip we programmed. The components needed for the circuit were listed in Chap. 1; if you purchased the components, you can quickly set up the test circuit. If not, the components are listed again at the end of this chapter; you will need the components to build the circuit.

The solderless breadboard

For those of us who have not dabbled in electronics very much, I want to describe the solderless breadboard (see Fig. 6.1) in detail. As the name implies, you can breadboard (assemble and connect) electronic components onto it without solder. The breadboard is reusable; you can change, modify, or remove circuitry components from the breadboard at any time. This makes it easy to correct any wiring errors. The solderless breadboard is an important item for constructing and testing circuits outlined in this book.

The style of breadboard on the left is available from any number of sources including RadioShack. The breadboard on the right is similar but provides a larger prototyping area.

If you wish to make any circuit permanent, you can transfer the components onto a standard printed-circuit board and solder it together with the foreknowledge that the circuit functions properly.

A partial cutaway of the top surface shows some of the internal structure of a board (Fig. 6.2). The holes on the board are plugs. When a wire or pin is inserted into the hole, it makes intimate contact with the metal connector strip inside. The holes are properly distanced so that integrated circuits and many other components can be plugged in. You connect components on the

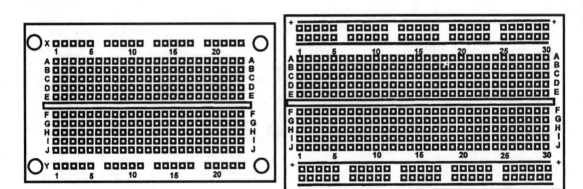

Figure 6.1   Top view of solderless breadboards.

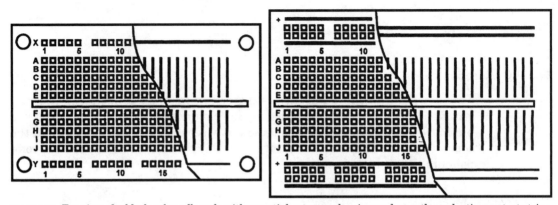

Figure 6.2   Top view of solderless breadboards with a partial cutaway showing underneath conductive contact strips.

board by using 22-gauge (solid or stranded) wire. I prefer to use stranded wire because it has greater flexibility; other people prefer solid wire because it's stiffer and easier to push into the breadboard hole.

The complete internal wiring structure of the solderless boards is shown in Fig. 6.3. The solderless breadboard on the left shows the X and Y rows that are typically used to supply power (Vcc) and ground connections to the circuit. The columns below the X row and above the Y row are used for mounting components. The solderless breadboard on the right has double rows located at the top and bottom. These are used to supply both Vcc and ground on each side of the breadboard.

Three schematics, one circuit

Figures 6.4, 6.5, and 6.6 are identical schematics of our test circuit. The 16F84 PIC microcontroller in the schematic is the microcontroller you programmed in either Chap. 4 or 5. I drew three schematics to help orient experimenters who may not be familiar with standard electrical drawings. Figure 6.4 shows

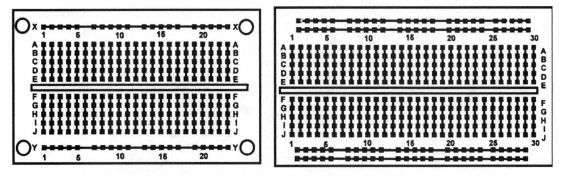

Figure 6.3 Top view of solderless breadboards detailing conductive strips.

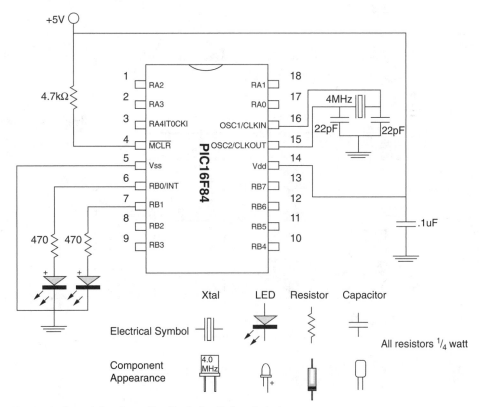

Figure 6.4 Isometric schematic of test circuit for `wink.bas` program.

how the PIC 16F84 microcontroller and components appear. There is a legend at the bottom that shows the electrical symbol and the typical appearance of the component. Figure 6.5 is a line drawing showing how the components appear mounted on one of the solderless breadboards. The writing on Fig. 6.5 points out each electrical component.

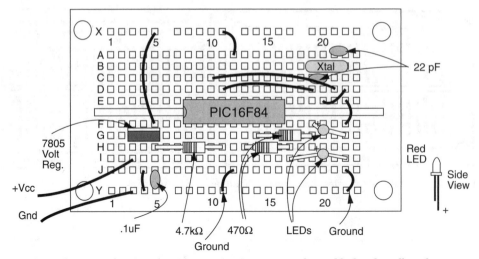

Figure 6.5 Isometric drawing showing test circuit constructed on solderless breadboard.

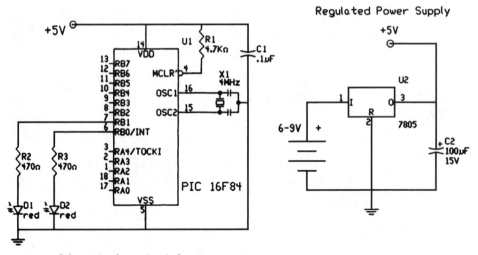

Figure 6.6 Schematic of test circuit for wink.bas program.

If you examine the placement of the components mounted on the solderless breadboard with its internal electrical wiring (Figs. 6.2 and 6.3), you can see how the components connect to one another and produce a circuit.

Figure 6.6 is the same schematic drawn as a standard electrical drawing with the pin numbers grouped and oriented to function. For the remainder of the book, standard electrical drawings will be used.

The schematic shows how minimal are the components needed to get your microcontroller up and running. Primarily you need a pull-up resistor on pin

Figure 6.7   Photograph of wink.bas circuit constructed on solderless breadboard.

4 (MCLR), a 4-MHz crystal with two (22-pF) capacitors and a 5-V power supply. *Note:* The 4-MHz crystal and two (22-pF) capacitors make up an oscillator that is required by the microcontroller. These three parts may be substituted with a 4-MHz ceramic resonator.

The two LEDs and the two resistors connected in series with each LED are the output. It allows us to see that the microcontroller and program are functioning properly.

Assemble the components as shown in the schematic (Fig. 6.5) onto the solderless breadboard. When you have finished, your work should appear as in Fig. 6.7.

Although the specifications sheet on the 16F84 states the microcontroller will operate on voltages from 2 to 6 V, I provided a regulated 5-V power supply for the circuit. The regulated power supply consists of a 7805 voltage regulator and two filter capacitors.

Wink

Apply power to the circuit. The LEDs connected to the chip will alternately turn on and off. Wink, ..., wink. Now you know how easy it is to program these microcontrollers and get them up and running.

Troubleshooting the circuit

There is not too much that can go wrong here. If the LEDs do not light, the first thing to check is the orientation of the LEDs. If they are put in backward, they will not light.

Next check your ground wires. See the jumper wires on the right-hand side of the solderless breadboard. They bring the ground up to the two 22-pF capacitors.

Check all your connections. Look back at Figs. 6.2 and 6.3 to see how the underlying conductive strips relate to the push in terminals on top of the board.

PIC Experimenter's Board and LCD Display

There are two optional tools you may want if you plan on experimenting with the PIC16F84 and microcontrollers in general. They are the PIC Experimenter's Board and LCD display. We will look at the LCD display first because a similar LCD display is incorporated into the PIC Experimenter's Board and what we say about the stand-alone LCD display is also true for the PIC Experimenter's Board LCD display.

One thing PIC microcontrollers lack is some type of display. With a display, the chip could show us how a program is running or what it is detecting. In addition a display would allow the microcontroller to output textual and numeric messages to the user.

To this end there are serial LCD displays on the market that only require a single microcontroller's I/O lines (pin) and a circuit ground. The particular LCD display we are using receives standard serial data (RS-232) at 300, 1200, 2400, and 9600 baud (Bd) (inverted or true). The LCD module is a two-line, 16-character visible display. The full display is actually two lines by 40 characters, but the additional 24 characters per line are off screen. We can use the PicBasic and PicBasic Pro `serout` command to communicate and output messages to the LCD display.

The PicBasic and PicBasic Pro compilers can send and receive serial information at 300, 1200, 2400, and 9600 Bd. Data are sent as 8 bits, no parity, and 1 stop bit. The serial mode may be set to be true or inverted. These data match the serial communication protocols required of the LCD display.

The LCD module has three wires: +5 V (red), GND (black or brown), and a serial in line (white). The baud rate may be set to 300, 1200, 2400, or 9600 by using a set of jumpers (J1, J2, and J3) on the back of the LCD display.

This first program prints the message "Hello World." The cursor (printing position) automatically moves from left to right. The schematic is shown in Fig. 6.8, and the LCD display is shown in Fig. 6.9.

```
'PicBasic program
'LCD test
pause 1000              'Wait 1 second for LCD to initialize
```

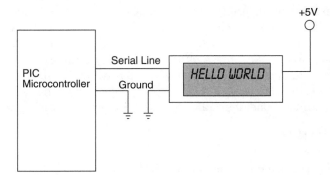

Figure 6.8 Schematic of LCD display test circuit.

Figure 6.9 Photograph of LCD display "Hello World."

```
start:
serout 1, t1200, (254,1)            'Clear screen
pause 40
serout 1, t1200, ("Hello World")    'Print message
pause 400
goto start
end
```

I kept this program small to show how easy it is to print a message on the LCD display. Here is the same program written for the PicBasic Pro compiler.

```
'PicBasic Pro program
'LCD test
pause 1000                            'Wait 1 second for LCD to initialize
start:
serout portb.1, 1, [254,1]            'Clear screen
pause 40
serout portb.1, 1, ["Hello World"]    'Print message
pause 400
goto start
end
```

Notice that, in line 5 of the program(s), `serout 1, t1200, (254,1)` is a command. The LCD module has eight common commands. All commands are prefixed with the decimal number 254. The LCD module will interpret any number following a 254 prefix as an instruction. Instead of decimal numbers, you may also use hexadecimal numbers, if you wish. So in hexadecimal the command becomes `serout 1, t1200, ($fe, $01)`. The following is a list of a few common commands. Remember all commands are prefixed with a 254 ($fe).

Code	Instruction
1	Clear screen.
2	Home position (move cursor top left of display).
16	Move cursor one character position left.
20	Move cursor one character position right.
24	Scroll display one character position left.
28	Scroll display one character position right.
192	Move cursor to first position on second line.

PIC Experimenter's Board

The PIC Experimenter's Board is a prefabricated developing board for prototyping circuits (see Fig. 6.10). The board allows easy access to all the I/O pins, port A (RA0–RA4), and port B (RB0–RB7) of the 16F84. The board may also be used with the 16F8X, 16C55X, 16C62X, 16C7X, and 16C8X family of 18-pin PIC microcontrollers.

Its 168-point solderless connection area allows for quick and easy access to all port A (RA0–RA4) and port B (RB0–RB7) I/O lines. There is an open 18-pin socket for inserting the microcontroller you are developing. The board includes an integrated 16 × 2 serial LCD display (optional backlight), which can be easily connected with one wire to any I/O line (or external source).

Use

The board can be powered by either an onboard 9-V battery or an ac/dc transformer. The power switch in the upper right turns power to the board on and off. The board includes a reset button, for resetting the microcontroller. The LCD has its own power switch, located directly above the LCD. If your LCD has a backlight, the backlight switch is located above the LCD power switch.

I will describe the prototyping section on the PIC Experimenter's Board, as I did with the solderless breadboards, and finish up the description by wiring a simple microcontroller LED project on the Experimenter's Board. The prototyping is located at the lower left corner of the PIC Experimenter's Board (see Fig. 6.11). There is an open 18-pin socket to hold the microcontroller being developed.

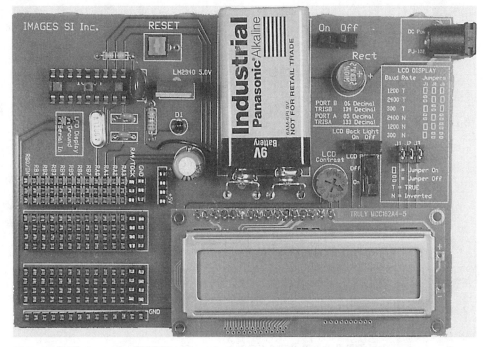

Figure 6.10 Photograph of PIC Experimenter's prototype developing board.

The prototyping area is similar in design and function to solderless bread-boards; see Fig. 6.12. You can breadboard (assemble and connect) electronic components and electronic circuits into the prototyping area without solder-ing. The prototyping area is reusable; you can change, modify, or remove cir-cuit components at any time. This makes it easy to correct any wiring errors.

A cutaway of the prototyping area is shown in Fig. 6.13. The square holes shown in the area are sockets. When a wire or pin is inserted into a hole, it makes electrical contact with the underlying metal strip. The holes are spaced so that integrated circuits and many other components can be plugged right in.

The internal electrical connection structure of the prototyping area is shown in Fig. 6.14.

Looking at Fig. 6.15, at the top of the prototyping area we see that the columns of bank 1 are labeled with the pin assignments from the 16F84. These columns are directly connected to those microcontroller pins. Connecting a wire or device to any of the three sockets in a particular column is electrically connecting that wire or device to that I/O pin of the 16F84.

Bank 2 provides 14 individual four-socket columns. The four sockets aligned in each individual column are electrically connected. The individual columns are separate electrically from one another.

Bank 3 is the same as bank 2.

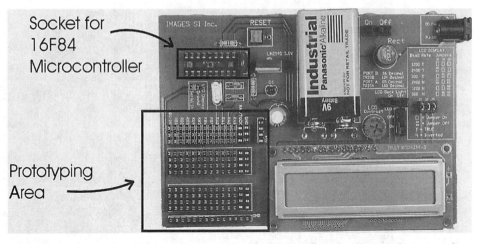

Figure 6.11   Photograph of PIC Experimenter's Board with breadboarding area and 18-pin socket highlighted.

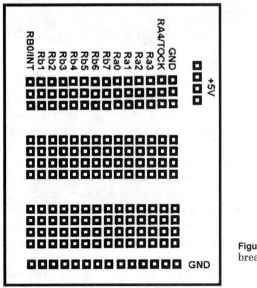

Figure 6.12   Diagram of the breadboard area.

The last row, labeled GND (ground), is electrically connected across the entire row. There are an additional three ground sockets at the top of bank 1.

A +5-V power is available from a four-socket column adjacent to bank 1.

Simple experiment

We shall wire a simple experiment to illustrate the use of the experimenter's prototyping area: blinking an LED. Yes, this is very similar to the wink pro-

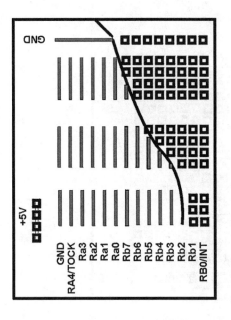

Figure 6.13 Cutaway view of breadboard area.

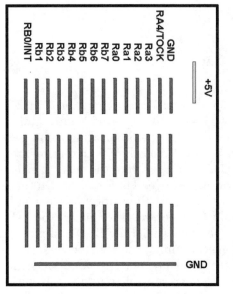

Figure 6.14 Underlying electrical connections of breadboard area.

gram, with the exception that we are only using one LED this time. The following are a small PicBasic program and PicBasic Pro program to blink an LED on pin RB1.

PicBasic program	**PicBasic Pro program**
start: high 1	start: high portb.1
pause 250	pause 250

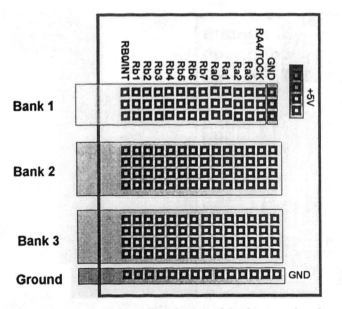

Figure 6.15 Diagram of breadboard area with banks, ground, and +5-V power supply highlighted.

```
low 1                    low portb.1
pause 250                pause 250
goto start               goto start
```

The complete schematic for this experiment is shown in Fig. 6.16. Aside from a programmed 16F84, we only need two other components: a 470-Ω, $\frac{1}{4}$-W resistor and a subminiature LED. All the other components needed to make the 16F84 work are already hardwired on the PIC Experimenter's Board.

The LED has two terminals, one longer than the other. The longer terminal on the LED is positive, shown in the legend of Fig. 6.17. On the schematic the LED appears as a diode. To wire this circuit, connect one lead of the $\frac{1}{4}$-W resistor into one of the RB1 sockets. Connect the other lead of the $\frac{1}{4}$-W resistor into a socket in bank 2. Take the positive lead of the LED and plug it into a socket in the same column as the one containing the resistor lead. Connect the opposite lead of the LED, and plug it into one of the ground sockets at the bottom.

Plug the programmed 16F84 microcontroller into the 18-pin socket on the PIC Experimenter's board, and turn on the power. The LED should begin blinking on for $\frac{1}{4}$ s, then off for a $\frac{1}{4}$ s. This on/off cycle (blinking) continually repeats.

Built-in LCD

The onboard LCD display combines a serial interface and a 2-line by 16-character display (see Fig. 6.18). The LCD display can be set to receive serial data at 300, 1200, or 2400 Bd (true or inverted, switch-selectable).

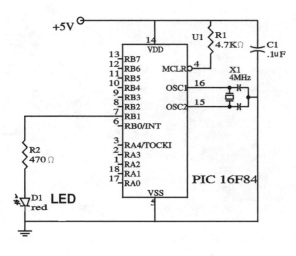

Figure 6.16 Schematic of blink circuit.

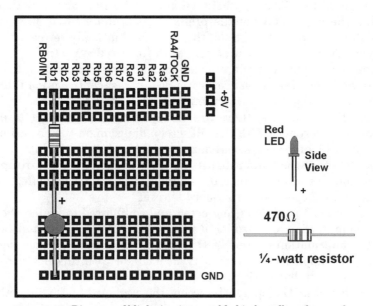

Figure 6.17 Diagram of blink circuit assembled in breadboard area of PIC Experimenter's Board.

To use the LCD, connect a jumper from the desired output pin on the microcontroller to the serial input. It is not necessary to connect a secondary ground line to the serial input ground unless the serial data are coming from a source off the Experimenter's Board.

Like the LCD module, the onboard LCD display has two operational modes: text and instruction. The default is text mode; data received via the serial input line appears on the screen. Send the string "Images" and "Images" will appear on the LCD. To input instructions to the LCD display, such as clear

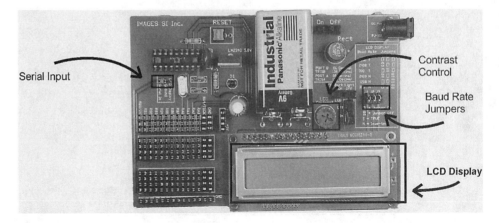

Figure 6.18 Photograph of PIC Experimenter's Board with LCD controls highlighted.

screen, go to line 2, etc., you must prefix the instruction with ASCII 254 (0xFE). The byte following the prefix is seen and treated as an instruction code. After the instruction code, the unit automatically returns to text mode. Every instruction code must be sent with its own 254 prefix.

If your LCD is backlit, you may adjust the backlight contrast to the optimal setting via the LCD contrast control. The contrast control is set fully clockwise (highest contrast) at the factory, but you can adjust the control by hand.

To set the baud rate, there are three sets of jumpers: J1 to J3. Set the jumpers in accordance with the silkscreen diagram on the Experimenter's PC Board. At all baud rates, serial data are received at 8 data bits, 1 stop bit, no parity. Note that the baud rate setting is only read once at start-up, so changing the jumpers while the module is active will not have any effect on the baud rate until the Experimenter's Board is reset.

Once the LCD module is connected and configured to match the baud rate of the computer/microcontroller, it will receive those transmitted data and display the information on the LCD display. For example, if you send "Hello," then "Hello" appears on the display. The cursor (printing position) automatically moves from left to right.

The onboard LCD display will accept the standard LCD instructions. A particular byte is identified as an instruction when it is preceded by an instruction prefix character, ASCII 254 (0xFE hex, 11111110 binary). The onboard LCD treats the byte immediately after the prefix as an instruction, then automatically returns to data mode. For example, the clear-screen instruction is ASCII 1. To clear the screen, send <254><1> (where the < > symbols mean single bytes set to these values, not text as typed from the keyboard). Notice this instruction code matches the instruction code for the serial LCD display module.

Instruction	Code (decimal)
Clear screen	1

Figure 6.19 Photograph of LCD display in self-test mode.

Home position (move cursor top left of display)	2
Move cursor one character position left	16
Move cursor one character position right	20
Scroll display one character position left	24
Scroll display one character position right	28
Set cursor position (DDRAM address)	128 + addr
Set point in character generator (CG) RAM	64 + addr

The onboard LCD also has a self-testing mode that will print the current baud rate as determined by the jumper settings and mode (true/inverted); see Fig. 6.19. To enter self-test mode, connect the serial in line to ground (for true) or +5 V (for inverted) upon LCD module start-up.

Note: If the serial input line is improperly connected for self-test mode, for instance connected to +5 V when jumpers are set for true mode, the LCD display will remain blank. The module stays in self-test mode as long as the serial input line is held either high (inverted mode) or low (true mode). LCD module may be exited from self-test mode on the fly by simply connecting the serial input line to a serial source.

When you print past the end of a line, the next 24 characters do not show up on the LCD screen. They are not lost; they are in an off-screen memory area. All alphanumeric LCD modules have 80 bytes of memory, arranged appropriately for a 2 × 40 screen. On LCDs with smaller screens (such as this 2 × 16), text printed past the end of a visible line goes into memory, but can't be seen on the screen. Use cursor-positioning instructions to print to a particular location on the display. Or deliberately print in off-screen memory to temporarily hide text, then send scroll-left instructions to reveal it.

Using the LCD: PicBasic and PicBasic Pro examples

Connect the serial input of the LCD to portb.0 of a PIC microcontroller. The following PicBasic program demonstrates sending data and commands to the LCD.

```
main: pause 1000                      'Wait for the LCD to start up
   serout 0, t2400, ($fe,$01)         'Clear the screen
```

```
pause 40
serout 0, t2400, ("Wherever you go")
serout 0, t2400, ($fe,$c0)              'Move to line 2
serout 0, t2400, ("there you are")
pause 1000                              'Wait 1 second
goto main                               'Do it again
```

The program clears the LCD and sends the message "Wherever you go there you are" at 2400 Bd (true mode), waits for 2 s, and then loops indefinitely (see Fig. 6.20). Note that in the above example, the control codes were written in hexadecimal (base-16). Hexadecimal is specified by a dollar sign ($) prefix. If you wish, the lines

```
serout 0, t2400,($fe,$01)    'Clear the screen
serout 0, t2400,($fe,$c0)    'Move to line 2
```

could also be written with decimal (base-10) notation as

```
serout 0, t2400,(254,1)      'Clear the screen
serout 0, t2400,(254,192)    'Move to line 2
```

Which notation you choose to use is a matter of preference. An equivalent program can be written with PicBasic Pro as follows:

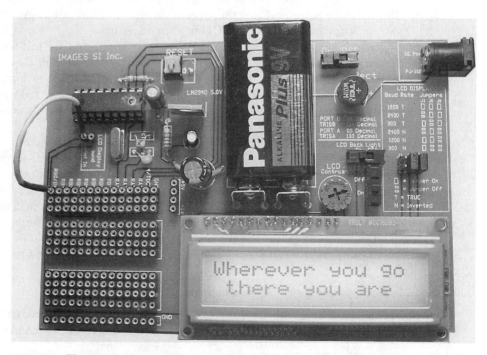

Figure 6.20 Photograph of message from pin RB0 of 16F84 microcontroller sent to onboard LCD display.

```
main: pause 1000                          'Wait for the LCD to start up
serout portb.0,0,[$fe,$01]                'Clear the screen
pause 40
serout portb.0,0,["Wherever you go"]
serout portb.0,0,[$fe,$c0]                'Move to line 2
serout portb.0,0,["there you are"]
pause 1000                                'Pause for a second
goto main                                 'Loop
```

Introduction to Binary and the PIC Microcontroller

The term *binary* means "based on 2," as in two numbers 0 and 1. It's also like an electric switch that has two values: on (1) and off (0).

The term *bit* is an acronym that stands for the term *binary digit*. A bit or binary digit can have two values, either 0 or 1. A *byte* is a digital expression (number) containing 8 bits.

Binary is important to computers and microcontrollers. The bit values of 0 and 1 are the only things a computer can read. Actually the computer or microcontroller can't really read, but it can sense voltage values. So a bit that is on 1 is represented by a positive voltage. Consequentially a bit 0 is off 0 and is represented as no voltage.

A single bit by itself is of little value, but start putting them together to make bytes (8 bits), words (16 bits, 32 bits, 64 bits, 128 bits), and so on, and we can make the computers perform mathematics, create word processors and spreadsheets, create a cyberspace (Internet), etc. All these applications are based on a bit.

To read or write to a port register requires understanding a little binary. When we read and write to any port, we use standard decimal numbers. However it's the binary equivalent of those decimal numbers that the PIC microcontroller sees and uses.

The 16F84 uses 8-bit port registers so we only need to concern ourselves with small 8-bit numbers and their decimal equivalents. Remember that an 8-bit number is called a byte. An 8-bit number can represent any decimal value between 0 and 255. When we write a decimal number into a register, the PIC microcontroller can only see the binary equivalent of that decimal number (byte) we wrote to the register. For us to understand what's happening inside the register, we need to be able to look at the binary equivalents of the decimal (byte) number also. Once we can do this, our ability to effectively and elegantly program the PIC microcontrollers is greatly enhanced.

Examine the binary number table at the top of the next page. It shows all the decimal and binary number equivalents for numbers 0 through 32. By using this information, the binary numbers from 32 to 255 can be extrapolated.

Each decimal number on the left side of the equals sign has its binary equivalent on the right side. So where we see a decimal number, the microcontroller will see the same number as a series of 8 *bits* (there are 8 bits to a byte).

Binary Number Table

0 = 00000000	16 = 00010000	32 = 00100000
1 = 00000001	17 = 00010001	.
2 = 00000010	18 = 00010010	.
3 = 00000011	19 = 00010011	.
4 = 00000100	20 = 00010100	64 = 01000000
5 = 00000101	21 = 00010101	.
6 = 00000110	22 = 00010110	.
7 = 00000111	23 = 00010111	.
8 = 00001000	24 = 00011000	128 = 10000000
9 = 00001001	25 = 00011001	.
10 = 00001010	26 = 00011010	.
11 = 00001011	27 = 00011011	.
12 = 00001100	28 = 00011100	255 = 11111111
13 = 00001101	29 = 00011101	
14 = 00001110	30 = 00011110	
15 = 00001111	31 = 00011111	

Figure 6.21 shows the relationship between a binary number and the two PIC microcontroller registers that control port b. Notice each register has eight open positions. This register can hold an 8-bit (1-byte) number. Let's look at the second binary number table below. Notice for each progression of the binary 1 to the left, the exponential power of 2 is increased by 1.

Second Binary Number Table

Bit no.	Decimal		Binary	Bit no.	Decimal		Binary
Bit 0	1	=	00000001	Bit 4	16	=	00010000
Bit 1	2	=	00000010	Bit 5	32	=	00100000
Bit 2	4	=	00000100	Bit 6	64	=	01000000
Bit 3	8	=	00001000	Bit 7	128	=	10000000

These are relevant numbers, because each progression to the left identifies another bit location and bit weight within the 8-bit byte.

For instance, suppose we wanted to write binary 1s at the RB6 and RB2 locations. To do so, we add their bit weights, in this case 64 (RB6) plus 4 (RB2), which equals 68. The binary equivalent of decimal number 68 is 01000100. If you push that number into the port B register, you will see that the binary 1s are in the RB6 and RB2 positions. Remember this—it is important.

The open TRISB register shown in Fig. 6.21 may be used to examine numbers placed in the TRISB. The port B register may be used to examine numbers placed at the port B register.

Notice the correlation between the register bit locations, bit weights, and

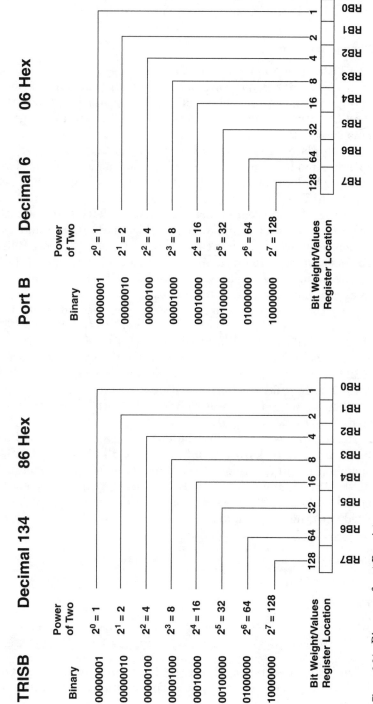

Figure 6.21 Diagram of port B registers.

67

port B I/O pins. This correspondence between the bit number, bit weight, and the I/O line is used to program and control the port.

Using the TRIS and port registers

The TRIS (tri-state enable) register is a 1-byte (8-bit) programmable register on the PIC 16F84 that controls whether a particular I/O pin is configured as an input or output pin. There is a TRIS register for each port. TRISA controls the I/O status for the pins on port A, and TRISB controls the I/O status for the pins on port B.

If one places a binary 0 at a bit location in TRISB for port B, the corresponding pin location on port B will become an output pin. If one places a binary 1 at a bit location in the TRISB, the corresponding pin on port B becomes an input pin. The TRISB data memory address for port B is 134 (or 86h in hex).

After port B has been configured using the TRISB register, the user can read or write to the port, using a port B address (decimal number 6).

Here is an example. Suppose we want to make all port B lines output lines. To do so, we need to put a binary 0 in each bit position in the TRISB register. So the number we would write into the register is decimal 0. Now all our I/O lines are configured as output lines.

If we connect an LED to each output line, we can see a visual indication of any number we write to port B. If we want to turn on the LEDs connected to RB2 and RB6, we need to place a binary 1 at each bit position on port B register. To accomplish this, we look at the bit weights associated with each line. RB2 has a bit weight of 4, and RB6 has a bit weight of 64. We add these numbers (4 + 64 = 68) and write that number into the port B register.

When we write the number 68 into the port B register, the LEDs connected to RB2 and RB6 will light.

To configure port A, we use the TRISA register, decimal address 133 (see Fig. 6.22). On port A, however, only the first 5 bits of the TRISA and the corresponding I/O lines (RA0–RA4) are available for use. Examine the I/O pin-out on the 16F84, and you will find there are only five I/O pins (RA0–RA4) corresponding to port A. These pins are configured using the TRISA register and used with the port A address.

Register	Memory location, hexadecimal	Memory location, decimal
Port A	05h	5
Port B	06h	6
TRISA	85h	133
TRISB	86h	134

On power up and reset, all the I/O pins of port B and port A are initialized (configured) as input pins. We can change this configuration with our program.

Port A

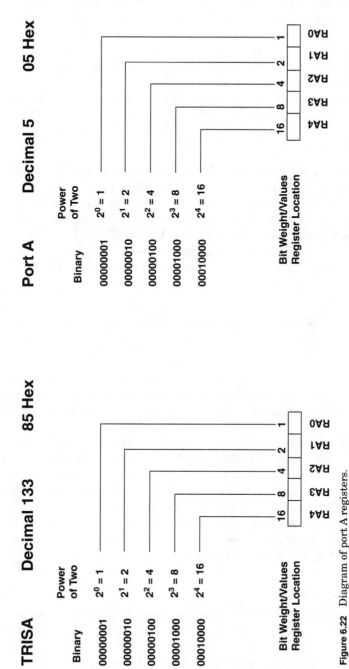

Figure 6.22 Diagram of port A registers.

Here's another example. Let's configure port B so that bit 0 (RB0) is an input pin and all other pins are output lines. To place binary 0s and 1 in the proper bit location, we use the bit weights shown in the binary number table. For instance, to turn bit 0 on (1) and all other bits off (0), we would write the decimal number 1 into TRISB for port B.

Depending upon which PicBasic compiler is used, the commands are a little different. For the PicBasic compiler, the command to write to a register is the `poke` command. The program line to write the decimal value 1 into the TRISB register will look like this:

```
poke 134,1
```

The number after the `poke` command is the memory address that the command will write to, in this case 134. The number 134 is the memory address of the TRISB for port B. The next number, separated by a comma, is the value we want to write in that memory address. In this case it's the number 1.

For the PicBasic Pro compiler, the TRISB and TRISA registers are already predefined. Thus when the compiler sees TRISB, it accesses the proper memory (134) location. So the equivalent command for the PicBasic Pro is

```
TRISB = 1
```

Look at the binary equivalent of the decimal number 1:

$$0\ 0\ 0\ 0\ 0\ 0\ 0\ 1$$

Mentally place each 1 and 0 into the TRISB register locations shown in Fig. 6.21. See how the 1 fits into the bit 0 place, making that corresponding line an input line, while all other bit locations have a 0 written in them, making them output lines.

So by poking (writing) this location with a decimal number that represents a binary number containing the proper sequence of bits (0s and 1s), we can configure any pin in the port to be either an output or an input in any combination we might require. In addition, we can change the configuration of the port "on the fly" as the program is running.

To summarize, writing a binary 1 into the TRIS register turns that corresponding bit/pin on the port to an input pin. Likewise, poking a binary 0 will turn the bit into an output.

Accessing the ports for output

Once the port lines have been configured (input or output) using the TRIS register, we can start using it. To output a binary number at the port, simply write the number to the port, using the `poke` (PicBasic) or `trisx.x` (PicBasic Pro) command. The binary equivalent of the decimal number will be outputted, as shown in our first example. To output a high signal on RB3 using the PicBasic compiler, we could use this command:

* Capacitors connected to Crystals are 22pF

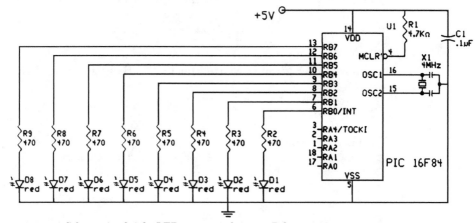

Figure 6.23 Schematic of eight LEDs connected to port B for counting program.

```
poke 6, 8
```

where 6 is the memory address for port B and 8 is the decimal equivalent of the binary number (00001000) we want to output.

For the PicBasic Pro compiler, the equivalent command is

```
output portb.3 = 1
```

Counting program

To illustrate many of these concepts, I have written a simple basic program. The schematic for the program is shown in Fig. 6.23. It is a binary counting program that will light eight LEDs connected to port B's eight output lines.

The counting program will light the LEDs in the sequence shown in the binary number table. Each binary 1 in a number the table will be represented with a lit LED. Every 250 milliseconds (ms) ($^1/_4$ s), the count increments. After reaching the binary number 255 (the maximum value of a byte), the sequence repeats, starting from zero.

Counting in binary by 1

The following program is written for the PicBasic compiler.

```
'Program binary counting
'Initialize variables
symbol trisb = 134          'Assign TRISB of port b to decimal value of 134
symbol portb = 6            'Assign port b to decimal value of 6
'Initialize port(s)
poke trisb,0                'Set port b pins to output
loop:
```

```
for b0 = 0 to 255
poke portb, b0          'Place count at port b to light LEDs
pause 250               'Pause 1/4 s or it's too fast to see
next b0                 'Next counter value
goto loop               'Start over again
'End
```

The following program is written for the PicBasic Pro compiler.

```
'Program binary counting
'Initialize variables
ct var byte             'Counting variable
'Initialize port
trisb = 0               'Set port b pins to output
loop:
for ct = 0 to 255       'Counter
portb = ct              'Place counter on port b to light LEDs
pause 250               'Pause 1/4 s
next ct                 'Next counter value
goto loop               'Start over again
'End
```

Input

The ability of our microcontroller to read the electrical status of its pin(s) allows the microcontroller to see the outside world. The line (pin) status may represent a switch, sensor, or electrical information from another circuit or computer.

The button command

The PicBasic compiler comes equipped with a simple command to read the electrical status of a pin, called the button command. The button command, while useful, has a few limitations. One limitation of this command is that it may only be used with the eight pins that make up port B. The I/O pins available on port A cannot be read with the button command. Another limitation is that you cannot read multiple port pin inputs at once, only one pin at a time.

We will overcome these button command limitations using the peek command. But for the time being, let's use and understand the button command.

As the name implies, the button command is made to read the status of an electrical button switch connected to a port B pin. Figure 6.24 shows two basic switch schematics, labeled A and B, of a simple switch connected to an I/O pin.

The button command structure is as follows:

```
button pin, down, delay, rate, var, action, label
```

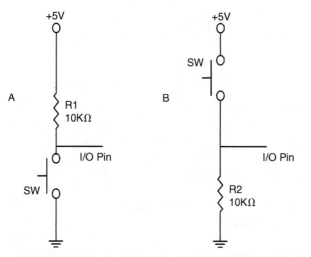

Figure 6.24 Schematic of electric switches suitable for use with PIC microcontrollers.

Pin   Pin number (0–7), port B.

Down   State of pin when button is pressed (0 or 1).

Delay   Cycle count before auto repeat starts (0–255). If 0, no debounce or auto-repeat is performed. If 255, debounce, but no auto-repeat is performed.

Rate   Auto-repeat rate (0–255).

Var   Byte variable used for delay/repeat countdown. Should be initialized to 0 prior to use.

Action   State of button to perform `goto` (0 if not pressed, 1 if pressed).

Label   Execution resumes at this label if Action is true.

Let's take another look at the switch schematic in Fig. 6.24 before we start using the button switch. Let's visualize how the switches affect the I/O pin electrically.

The switch labeled A in Fig. 6.24 connects the I/O pin to a +5-V power supply through a 10,000-Ω resistor. With the switch open, the electrical status of the I/O pin is kept high (binary 1). When the switch is closed, the I/O pin connects to ground, and the status of the I/O pin is brought low (binary 0).

The switch labeled B in Fig. 6.24 has an electrical function opposite the switch labeled A. In this case, when the switch is open, the I/O pin is connected to ground, keeping the I/O pin low (binary 0). When the switch is closed, the I/O pin is brought high (binary 1).

In place of a switch, we can substitute an electric signal, high or low, that can also be read using the `button` command.

Typically the `button` command is used inside a program loop, where the program is looking for a change of state (switch closure). When the state of the I/O pin (line) matches the state defined in the Down parameter, the program execution jumps out of the loop to the *label* portion of the program.

A `button` example

If we want to read the status of a switch of I/O pin 7, here is a command we will use in the next program.

```
button 7, 0,254,0,b1,1,loop
```

The next program is similar to the previous program 3, inasmuch as it performs a binary counting. However, since we are using PB7 (pin 7) as an input, and not an output, we lose its bit weight in the number we can output to port B. The bit weight for pin 7 is 128. So without pin 7 we can only display numbers up to decimal number 127 (255 − 128 = 127). This is reflected in the first loop (pin7/bit 7 = 128).

The program contains two loops. The first loop counts to 127, and the current number's binary equivalent is reflected by the Lite LEDs connected to port B. The loop continues to count as long as the switch SW1 remains open.

When SW1 is closed, the `button` command jumps out of loop 1 into loop 2. Loop 2 is a noncounting loop where the program remains until switch SW1 is reopened. You can switch back and forth between counting and noncounting states. Figure 6.25 is a schematic of our button test circuit.

The following program is written for the PicBasic compiler.

```
'Program for PicBasic compiler
symbol trisb = 134          'Set TRISB to 134
symbol portb = 6            'Set port b to 6
'Initialize Port(s)
poke trisb,128              'Set port b pins (1-6 output), pin 7 input
```

* Capacitors connected to crystals are 22pF.

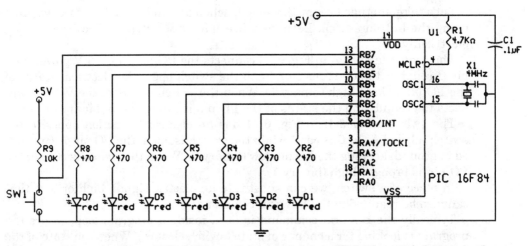

Figure 6.25 Schematic of seven LEDs and one switch connected to port B for the switch detection and counting program.

```
label 1: b1s = 0                    'Set button variable to 0
loop1:                              'Counting loop
for b0 = 0 to 127
poke portb, b0                      'Place b0 value at port to light LEDs
pause 250                           'Pause counting or it's too fast to see
button 7,0,254,0,b1,1,label2        'Check button status; if closed, jump
next b0                             'Next b0 value
goto loop1
label12: b1=0                       'Set button variable to 0
loop2:                              'Second loop not counting
poke portb,0                        'Turn off all LEDs
button 7,1,254,0,b1,1,label1        'Check button status; if open, jump back
goto loop2
```

When the program is run, it begins counting. When the switch is closed, all the LEDs will turn off, and it stops counting. Open the switch, and the counting resumes, starting from 0.

```
'Program for PicBasic Pro compiler
ct var byte
c1 var byte
'Initialize port(s)
trisb = 128                         'Set port b pins (1-6 output), pin 7 input
label1: c1=0                        'Set button variable to 0
loop1:                              'Counting loop
For ct = 0 to 127
portb = ct                          'Place ct value at port to light LEDs
pause 250                           'Pause counting or it's too fast to see
button 7,0,254,0,c1,1,label2        'Check button status; if closed, jump
next ct                             'Next counting value
goto loop1
label2: c1=0                        'Set button variable to 0
loop2:                              'Second loop not counting
portb =0                            'Turn off all LEDs
button 7,1,254,0,c1,1,label1        'Check button status; if open, jump back
goto loop2
```

peek

The peek command can only be used with the PicBasic compiler. We can also use the peek command to check the status of any input line. The advantages of the peek command are as follows. Using peek, we can read the five I/O lines of port A (or the eight I/O lines of port B) at once. This increases the versatility of the PIC chip and allows our program to be more concise (less convoluted), shorter, and easier to read.

To emphasize these points, let's rewrite our last programs, using the peek command. This program uses the same schematic.

```
'PicBasic program that uses the peek command
'Initialize port(s)
symbol trisb = 134         'Set TRISB to 134
symbol portb = 6           'Set port b to 6
poke trisb,128             'Set port b pins (1..6) output, pin 7 input
loop1:                     'Counting loop
for b0 = 0 to 127
poke portb, b0             'Place b0 value at port to light LEDs
pause 250                  'Pause counting or it's too fast to see
peek portb,b0              'Check button status
if bit7 = 0 then loop2     'If sw1 is closed, jump to loop2
next b0                    'Next b0 value
goto loop1
loop2:                     'Second loop not counting
pke portb,0                'Turn off all LEDs
peek portb,b0              'Check button status; if open, jump back
if bit7 = 1 then loop1     'If sw1 is open, jump to loop 1
goto loop2
```

The variable b0 is performing double duty. First it is holding our current counting numbers 0 through 127. The numbers 0 to 127 require 7 bits of the variable b0 (bit 0 through bit 6). This leaves the eighth bit (bit 7) available for use. We use bit 7 to check the status of the switch. If it's open, its value will be a binary 1; if it's closed, it's equal to binary 0.

The command peek is followed by a memory address, then a comma, then a storage variable.

```
peek address, var
```

As its name implies, the peek command allows one to view (or peek at) the contents of a specified memory address. Typically the memory address "peeked at" is one of the PIC microcontroller's registers. The "peeked" value is stored in a variable var defined in the command.

In this program we peeked at the one input line on port B:

```
peek portb,b0
```

The peek command can read an entire byte (8 bits) at once. Or as in this case, only the upper bit (bit 7) of the peeked value is relevant. (The rest of the bits are holding our counting number that's outputted to the LEDs).

peek and PicBasic Pro

When you are using the PicBasic Pro compiler, it is recommended *not* to use the peek command. Fortunately there is an easy work-around to the peek command. We simply assign a variable to the port we wish to peek at.

```
var = portb
```

The value placed in the variable var is our peek value.

```
'PicBasic Pro program that uses a peek equivalent command
ct var byte
cl var byte
'Initialize port(s)
trisb = 128              'Set port b pins (1..6) output, pin 7 input
loop1:                   'Counting loop
for ct = 0 to 127
portb = ct               'Place ct value at port to light LEDs
pause 250                'Pause counting or it's too fast to see
cl = portb               'Check button status
if cl.7 = 0 then loop2   'If sw1 is closed, jump to loop2
next ct                  'Next ct value
goto loop1
loop2:                   'Second loop not counting
portb = 0                'Turn off all LEDs
cl = portb               'Check button status; if open, jump back
if cl.7 = 1 then loop1   'If sw1 is open, jump to loop 1
goto loop2
```

Basic input and output commands

In our programs we directly wrote to the PIC microcontroller TRIS registers (A or B) and port registers. By doing so we were able to create input and output pins and then access them in our programs. There are other commands you can use to accomplish the same thing.

The PicBasic and PicBasic Pro compilers have two basic commands for making individual pins either input or output lines. The commands are input and output. Unfortunately these two basic commands only work on port B pins (0 to 7) for PicBasic. For PicBasic Pro, any port may be used. The command

```
input pin
```

makes the specified pin an input line. Only the pin number itself, that is, 0 to 7, is specified (i.e., *not* pin 0), for example,

```
input 2    'Makes pin2 (rb2) an input line.
```

The opposite of the input command is the output command. The command

```
output pin
```

makes the specified pin an output line. Only the pin number itself, that is, 0 to 7, is specified (i.e., *not* pin 0), for example,

```
output 0    'Makes port b, pin 0 (rb0) an output
```

The above examples are intended for use with either PicBasic or PicBasic Pro.

The PicBasic Pro has an additional command structure that can be used with both the `input` and `output` commands. This allows one to make input and output pins on other ports besides port B. This is accomplished by specifying the port and the pin.

For instance, to access port A, pin 2, you use the following format:

```
porta.2
```

To use this in a command:

```
input porta.2     'Make port a, pin 2 an input
output porta.3     'Make port a, pin 3 an output
```

Servomotors

Servomotors (see Fig. 6.26) are used in many radio-controlled model airplanes, cars, boats, and helicopters. Because of this large hobbyist market, servomotors are readily available in a number of stock sizes. Servomotors are used in a few of our robots.

Primarily, servomotors are geared dc motors with a positional feedback control that allows the rotor to be positioned accurately. The specifications state that the shaft can be positioned through a minimum of 90° (±45°). In reality we can extend this range closer to 180° (±90°) by adjusting the positional control signal.

There are three wire leads to a servomotor. Two leads are for power +5 V and GND. The third lead feeds a position control signal to the motor. The position control signal is a single variable width pulse. The pulse can be varied from 1 to 2 ms. The width of the pulse controls the position of the servomotor shaft.

A 1-ms pulse rotates the shaft to the extreme counterclockwise (CCW) position (−45°). A 1.5-ms pulse places the shaft in a neutral midpoint position (0°). A 2-ms pulse rotates the shaft to the extreme CW position (+45°).

The pulse width is sent to the servomotor approximately 50 times per second (50 Hz). Figure 6.27 illustrates the relationship of pulse width to servomotor position.

In most of the robots that use servomotors, the servomotor must be positioned to its center location before being assembled into the robot. To center the servomotor, we build a simple circuit and PicBasic program. The circuit is shown in Fig. 6.28. The programs for the PicBasic and PicBasic Pro compilers follow:

```
'PicBasic program to center servomotor
start:
pulsout 0, 150     'Send pulse out on rb0
pause 18           'Delay needed to send pulse at 55 Hz
goto start         'Repeat
```

Figure 6.26 Photograph of a servomotor.

The following program is for the PicBasic Pro compiler.

```
'PicBasic Pro program to center a servomotor
start:
pulsout portb.0, 150   'Send pulse out on rb0
pause 18               'Delay needed to send pulse at 55 Hz
goto start             'Repeat
```

This centering program and circuit will be referred to in later parts of the book when servomotors are discussed.

Parts List

LCD serial display
Pic Experimenter's Board

Available from Images SI Inc. (see Suppliers at end of book).

Microcontroller (16F84)	$7.95
4.0-MHz Xtal	$2.50

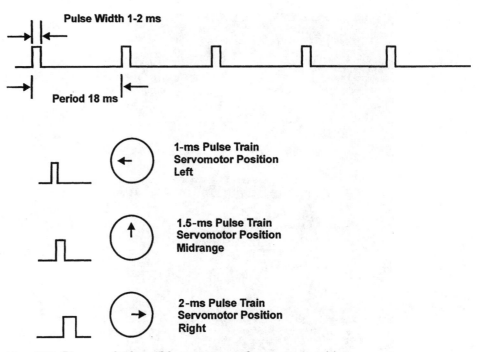

Figure 6.27 Diagram of pulse widths sent to control servomotor position.

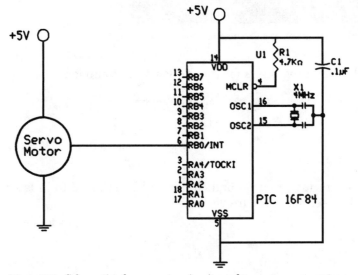

Figure 6.28 Schematic of servomotor circuit used to center servomotors.

(2) 22-pF capacitors

(1) Solderless breadboard   RadioShack PN# 276-175

(1) 0.1-μF capacitor   RadioShack PN# 272-1069

(2) Red LEDs   RadioShack PN# 276-208

(2) 470-Ω resistors*   RadioShack PN# 270-1115

(1) 4.7-kΩ resistor   RadioShack PN# 271-1126

(1) Voltage regulator (7805) RadioShack PN# 276-1770

(1) 9-V battery clip   RadioShack PN# 270-325

Available from RadioShack, Images SI Inc., Jameco Electronics, and JDR Microdevices (see Suppliers).

---

*These resistors are also available in 16-pin dip package.

Intelligence

Programming intelligence into a robot (or computer) is a difficult task and one that has not been very successful to date even when supercomputers are used. This is not to say that robots cannot be programmed to perform very useful, detailed, and difficult tasks; they are. Some tasks are impossible for humans to perform quickly and productively. For instance, imagine trying to solder 28 filament wires to a $^1/_4$-in square sliver of silicon in 2 s to make an integrated-circuit chip. It's not very likely that a human would be able to accomplish this task without a machine. But machine task performance, as impressive as it is, isn't intelligence.

Approaches to Building Intelligence

There are two schools of thought concerning the creation of intelligence in artificial systems. The first approach programs an expert system (top down); the second is a neural or behavior-based system (bottom up).

The expert system uses rules to guide the robot in task performance. Behavior-based programs create an "artificial" behavior in the robot that causes it to reflectively (automatically) perform the task required. Behaviors may be *programmed* (software) or may be *hardwired* into the robot. Behavior-based intelligence doesn't require a central processor, although such a system may have one.

Let's look at a practical programming problem and see how each approach differs. Suppose you worked for a company that designed a new robotic vacuum cleaner. The purpose of the robot is to vacuum the floor of a customer's home or apartment. Your job is to program the navigation system. The robot needs to move autonomously throughout the house. How would you go about programming the robot to accomplish navigation around the home so it could travel in and out of rooms without destroying the place?

Let's assume you first decide to try an expert navigation system. This approach uses brute-force programming and a lot of memory. You might begin by dividing the task of vacuuming the apartment or home into smaller tasks such as vacuuming individual rooms. You begin by programming into the robot's memory an electronic map (floor plan) of the home or area where the robot needs to vacuum. Then you map out each individual room and its contents. The robot must have the ability to measure its movement as it moves as well as compass direction to maintain its location integrity. Once this is accomplished, the robot must have an exact start location on the floor plan.

The robot's movement from the start position is measured and plotted on its internal floor plan map. Problems occur if an object is positioned differently or is out of place, such as a trash receptacle or chair that has been moved. In this situation the real world does not match the robot's internal map. Similar problems occur if new objects are left on the floor such as a bag, toy, or pet.

Even so, these obstacles would not present too much of a problem for an expert system. To compensate, a secondary collision detection subprogram could be written to detect, map, and go around an obstacle not existing on the internal map. The robot continues to move and vacuum the floor. Keep in mind that as the robot navigates around new obstacles, it's continually updating its internal map as it travels, to maintain its location integrity. These tasks are gobbling up computer time and memory.

The robot vacuum accomplished its task. Now suppose you want to share this robot or rent it. Now you have a problem. Each new house and every room in the new house would require its own electronic map. Although expert programming does work, it tends to be inflexible and not adaptive toward new or innovative situations.

Now let's try the other approach that uses behavior-based or bottom-up programming. Instead of programming internal maps, we program sensor responses and behavior-based algorithms (feedforward and feedback loops) for sensing and traveling around obstacles and avoiding getting stuck underneath furniture or trapped in corners. Without any internal map we allow the robot to travel and move around the house in a random manner. The idea is that while traveling in a haphazard manner, it will eventually make its way throughout the rooms, cleaning the floor as it goes. Because the robot travels randomly, it will take longer for the robot to vacuum the entire floor, and it may miss a spot here and there, but it gets the job done. Since this behavior-based type of robot vacuum isn't programmed for a particular house or room, it may be used in any house in any room at any time.

While our example is simple, it does illustrate the main differences between expert and behavior-based (neural) programming. But let's look at just one more example before we move on.

Expert systems typically have all the answers that the designers believe will be required by the system programmed into the system before it begins. It may store and categorize new information, but based on previously determined categories and existing knowledge. An example of this system could be a rock identification system. The robot examines unknown rocks based on known characteristics of rocks, such as color, hardness, scratchability, acid reaction tests, mass, etc. The expert system fails if it inadvertently picks up a piece of ice that melts to water during the tests. Well, it fails as long as the designer(s) never anticipated the robot picking up a piece of ice by mistake and made allowances for it.

Neural (behavior-based) systems are not programmed and are more adaptive, as shown in the previous example. But is a neural system suitable for this task of rock identification? Probably not! There are instances in which expert systems are the method of choice. One shouldn't blindly assume one system is better than the other in all cases.

To date, behavior-based robots are more successful at task accomplishments such as traveling over unfamiliar and rough terrain than are programmed robots. (Other neural-based intelligence includes speech recognition, artificial vision, speech generation, complex analysis of stock market data, and life insurance policies.)

Where's the Intelligence?

Behavior-based systems at their most basic level are neural reflex actions, so where's the intelligence in that? However, true behavior-based systems, when layered on top of one another, generate what appears to be (meaning to us homo sapiens) intelligence actions. This is not a consciousness mind, which is a whole other category of intelligence, but the layer behavior-based circuits mimic intelligent actions quite convincingly.

Layered Behavioral Responses

Let's layer a few behavioral responses on top of one another to see how intelligence behavior emerges. This particular robot is a modified "photovore." It will use a number of standard photoresistors as sensors.

Layer 1 is a simple on and off system. It uses a single photoresistor to read ambient light intensity. In darkness the system turns itself off and shuts down all electric power to the robot. When the ambient light increases to a low threshold, the system turns itself on and the robot travels forward slowly.

Layer 2 is a two-photoresistor sensor. It determines in which direction the light intensity is greater. These sensors steer the robot in the direction of the greatest light intensity.

Layer 3 is a single-photoresistor sensor. Under high-intensity light it shuts down the robot's drive system and allows the robot to bathe in strong-intensity light.

An outsider who didn't know how this robot was wired would observe the following behavior. At night the robot sleeps. At dawn it begins to travel, looking for a bright light source (food). When it finds a sufficiently bright light source, it stops to feed, recharging its batteries through solar panels.

So our simple photovore robot exhibits three (dare we say intelligent?) behaviors—sleep, searching or hunting, and feeding. That's not bad for a handful of components and some neural glue.

Behavior-Based Robotics

Behavior-based programs and robotics are not new concepts. Seminal work has been written and experiments carried out since the 1940s. In the 1940s neural networks and behavior-based robotics were hardwired electrical components.

In the 1940s Dr. W. Grey Walter built two turtlelike mobile robots that exhibit complex behavior using a few electrical neurons. The behavior generated was at the time called robotic reflexes. Today this behavior is more accurately described as layered neural architecture.

In the 1980s Valentino Braitenberg wrote a book entitled *Vehicles—Experiments in Synthetic Psychology* in which he described complex behavior emerging from the use of a few artificial neurons.

Rodney Brooks, head of MIT's Artificial Intelligence Laboratory, is a leader in the field of subsumption architecture, which again is behavior-based and neural.

Mark Tilden, creator of the nervous network technology, which again is reflex-based, doesn't program strategies such as walking into his biomorphic robots. Instead he creates a nervous network whose desired state creates a walking gait.

What these scientists have discovered is that neural behavior-based architecture offers unique advantages over standard-based expert programming.

In this book we will build a few behavior-based robots, using the PIC microcontroller extensively in their construction.

8

Walter's Turtle

Behavior-Based Robotics

Behavior-based robotics were first built in the 1940s. At that time these robots were described as exhibiting reflexive behavior. This is identical to the neural-based approach to implementing intelligence in robots, as outlined in Chap. 7.

William Grey Walter—Robotics Pioneer

The first pioneer in the bottom-up approach to robotics is William Grey Walter. William Grey Walter was born in Kansas City, Missouri, in the year 1910. When he was 5, his family moved to England. He attended school in the United Kingdom and graduated from King's College, Cambridge, in 1931. After graduation he began doing basic neurophysiological research in hospitals.

Early in his career he found interest in the work of the famous Russian psychologist Ivan Pavlov. Do you remember from your high school science classes the famous "Pavlov's dogs" stimulus-response experiment? In case you forgot, Pavlov rang a bell just before providing food for dogs. After a while the dogs became conditioned to salivate just by hearing the bell.

Another contemporary of Walter, Hans Berger, invented the EEG machine. When Walter visited Berger's laboratory, he saw refinements he could make to Berger's EEG machine. In doing so, the sensitivity of the EEG machine was improved, and new EEG rhythms below 10 Hz could be observed in the human brain.

Walter's studies of the human brain led him to study the neural network structures in the brain. The vast complexities of the biological networks were too overwhelming to map accurately or replicate. Soon he began working with individual neurons and the electrical equivalent of a biological neuron. He wondered what type of behavior could be gathered with using just a few neurons.

To answer this question, in 1948 Walter built a small three-wheel mobile robot. The mobile robot measured 12 in high and about 18 in long. What is fascinating about this robot is that by using just two electrical neurons, the robot exhibited interesting and complex behaviors. The first two robots were affectionately named Elmer and Elsie (*electrome*chanical *r*obot, *l*ight *s*ensitive). Walter later renamed the style of robots Machina Speculatrix after observing the complex behavior they exhibited.

In the early 1940s transistors had not been invented, so the electronic neurons in this robot were constructed by using vacuum tubes. Vacuum tubes consume considerably greater power than semiconductors do, so the original turtle robots were fitted with large rechargeable batteries.

The robot's reflex or nervous system consisted of two sensors connected to two neurons. One sensor was a light-sensitive resistor, and the other sensor was a bump switch connected to the robot's outer housing.

The three wheels of the robot are in a triangle configuration. The front wheel had a motorized steering assembly that could rotate a full 360° in one direction. In addition, the front wheel contained a drive motor for propulsion. Since the steering could continually rotate a full 360°, the drive motor's electric power came through slip rings mounted on the wheel's shaft.

A photosensitive resistor was mounted onto the shaft of the front wheel steering-drive assembly. This ensured that the photosensitive resistor was always facing in the direction in which the robot was moving.

Four Modes of Operation

While primarily a photovore (light-seeking) type of robot, the robot exhibited four modes of operation. It should be mentioned that the robot's steering motor and drive motor were usually active during the robot's operation.

Search. Ambient environment at a low light level or darkness. Robot's responses, steering motor on full speed, drive motor on $1/_2$ speed.

Move. Found light. Robot's responses, steering motor off, drive motor full speed.

Dazzle. Bright light. Robot's responses, steering $1/_2$ speed, drive motor reversed.

Touch. Hit obstacle. Robot's response, steering full speed, reverse drive motor.

Observed Behavior

In the 1950s Walter wrote two *Scientific American* articles ("An Imitation of Life," May 1950; "A Machine That Learns," August 1951) and a book titled *The Living Brain* (Norton, New York, 1963). The interaction between the neural system and the environment generated unexpected and complex behaviors.

In one experiment Walter built a hutch, where Elsie could enter and recharge its battery. The hutch was equipped with a small light that would draw the robot to it as its batteries ran down. The robot would enter the hutch, and its battery would automatically be recharged. Once the battery recharged, the robot would leave the hutch to search for new light sources.

In another experiment he fixed small lamps on each tortoise shell. The robots developed an interaction that to an observer appears as a kind of social behavior. The robots danced around each other, at times attracted and then repelled, reminding him of a robotic mating ritual or territorial marking behavior.

Building a Walter Tortoise

We can imitate most functions in Walter's famous tortoise. My adaptation of Walter's tortoise is shown in Fig. 8.1. To fabricate the chassis, we need to do a little metalwork. Working metal is made a lot easier with a few tools such as a center punch, hand shears, nibbler, drill, vise, and hammer (see Fig. 8.2).

Center punch: Used to make a dimple in sheet metal to facilitate drilling. Without the dimple, the drill is more likely to "walk" off the drill mark. Hold the tip of the center punch in the center of the hole you need to drill. Hit the center punch sharply with a hammer to make a small dimple in the material.

Shears: Used to cut sheet metal. I would advise purchasing 8- to 14-in metal shears. Use as a scissors to cut metal.

Nibbler: Used to remove (nibble) small bits of metal from sheet and nibble cutouts and square holes in light-gauge sheet metal. Note RadioShack sells an inexpensive nibbler.

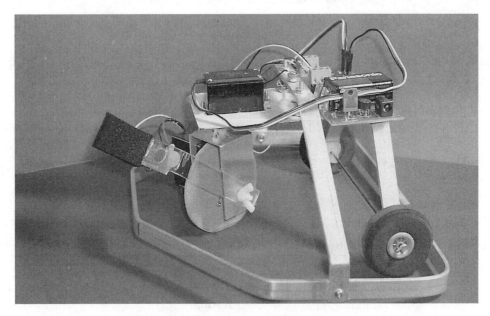

Figure 8.1 Adaptation of Walter's turtle robot.

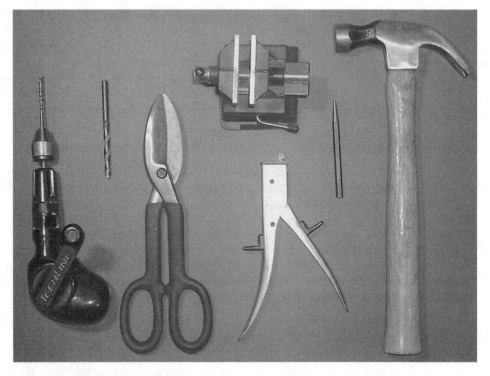

Figure 8.2 A few sheet metal tools.

Vise: Used to hold metal for drilling and bending.

Drill and hammer. Self-explanatory.

A well-stocked hardware store will carry the simple metalworking tools outlined. Most will also carry the light-gauge sheet metal and aluminum bar materials needed to make the chassis.

I built the chassis out of ($\frac{1}{8}$- $\times$ $\frac{1}{2}$-in) aluminum rectangle bar and 22- to 24-gauge stainless steel sheet metal. Stainless steel is harder to work with than cold rolled steel (CRS). And CRS is harder to work with than sheet aluminum. If I were to do this project over, I would use aluminum extensively because it is easier to work with than CRS or stainless steel.

Drive and Steering Motors

The robot uses servomotors for both the drive and steering. The drive servomotor is a HiTec HS-425BB 51-oz torque servomotor (see Fig. 8.3). The HS-425BB servomotor is modified for continuous rotation. For steering the robot I used a less expensive HiTec HS-322 42-oz torque servomotor (unmodified). Before we go into the robot fabrication, we must first modify the HS-425BB servomotor for continuous rotation.

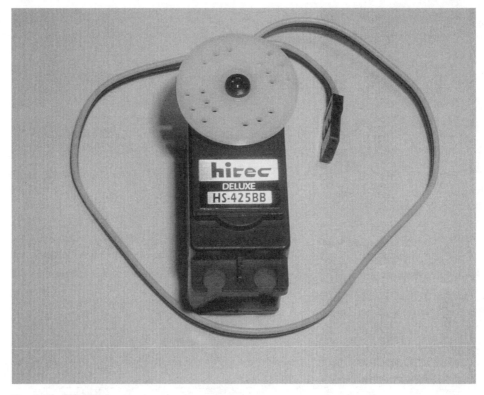

Figure 8.3 HS-425 servomotor.

Modifying the HS-425BB Servomotor

I chose the HS-425BB servomotor because I found it to be the easiest servo-motor to modify for continuous rotation. To create a continuous rotation ser-vomotor, it is necessary to mechanically disconnect the internal potentiometer from the output gear.

First remove the four back screws that hold the servomotor together (see Fig. 8-4). Keep the servomotor horn attached to the front of the servomotor. Once the screws are removed, gently pull off the front cover of the servomotor. The output gear will stay attached to the front cover, separating from the shaft of the potentiometer left in the servomotor's case (see Fig. 8.5). Sometimes the idler gear will fall out. Don't panic; it's easy enough to put back in position when you reassemble the servomotor.

Next remove the plastic clip from the servomotor shaft (see Fig. 8.6). With the plastic clip removed, the shaft of the potentiometer will no longer follow the rotation of the output gear. Align the potentiometer shaft so that the flat sides of the shaft are parallel to the long sides of the servomotor case (see Fig. 8.7).

Take off the front cover of the servomotor, and remove the center screw hold-ing the servomotor horn and output gear (see Fig. 8.8). The output gear is

Figure 8.4 Removing screws from back of servomotor case.

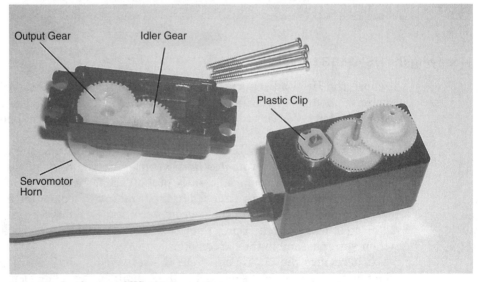

Figure 8.5 Inside view of HS-425 servomotor.

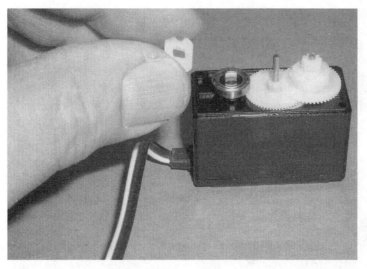

Figure 8.6   Removing plastic clip.

Figure 8.7   Top view of servomotor gears with plastic clip removed.

shown in Fig. 8.9. Remove the bearing from the output gear (see Fig. 8.10). The bearing needs to be removed so that you can cut away the stop tab from the gear. Use a hobby knife or miniature saw to cut away the stop tab. When you are finished cutting off the tab, check that the cut surfaces are smooth. If not, use a file to smooth out the surfaces.

Next remount the bearing onto the gear (see Fig. 8.11). Reassemble the idler and output gears onto the servomotor's gear train in the case (see Figs. 8.12 and 8.13). Now fit on the servomotor cover, and reattach the cover, using the four screws.

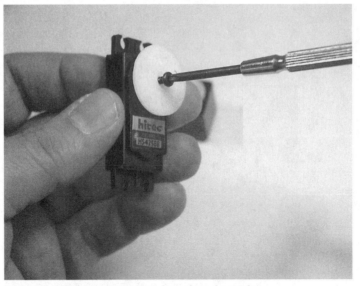

Figure 8.8 Removing servomotor horn from front of case.

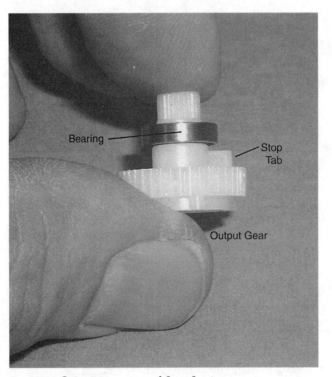

Figure 8.9 Output gear removed from front case.

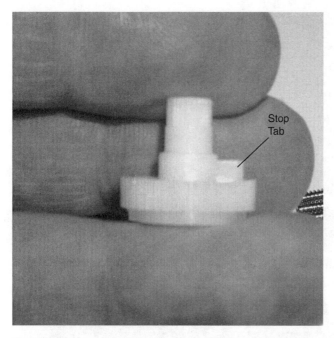

Figure 8.10 Stop tab on output that must be removed.

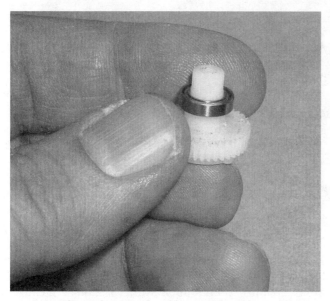

Figure 8.11 Stop tab removed and bearing placed back on gear.

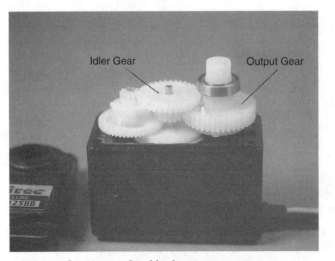

Figure 8.12 Output gear fitted back onto servomotor.

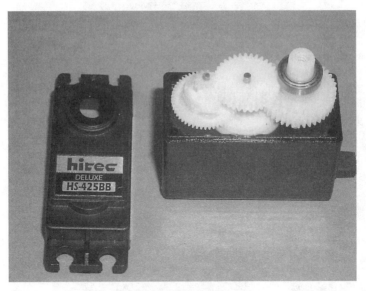

Figure 8.13 Ready for reassembly of servomotor.

The output shaft of the servomotor is now free to rotate continuously. A pulse width of 1 ms sent 50 to 60 times per second (Hz) will cause the servomotor to rotate in one direction. A pulse width of 2 ms will cause the servomotor to turn in the opposite direction.

There are two ways we can stop the servomotor from rotating. The first method is to simply stop sending pulses to the servomotor. The second method is a little trickier. A pulse width of approximately 1.5 ms will stop the servomotor. The exact pulse width for each servomotor must be determined experimentally. The exact pulse width required is based upon the position of the static potentiometer shaft inside the servomotor. If you followed the directions provided, it should be about 1.5 ms. To find the exact pulse width to stop the servomotor, you have two options. The first is to keep manually adjusting the pulse width until you find the correct pulse width. As you approach the pulse width needed to stop the servomotor, you will notice that the rotational speed of the servomotor will slow down. You can use this as a feature to create a speed control, if you wish.

The second option is to look at the servomotor circuit described in Chap. 14 (see Fig. 14.11). This simple circuit allows you to quickly find the correct pulse width.

Sheet Metal Fabrication

There are three pieces of sheet metal one needs to fabricate.

The U bracket, shown in Fig. 8.14, holds the front wheel and drive servomotor. The U bracket may be fabricated from 22-gauge 1.25- × 5-in aluminum sheet metal. I would recommend purchasing the U bracket (see Parts List) because the cutting required for this fabrication is extensive and precise.

The U bracket mounts the drive servomotor (see Fig. 8.15). In addition, on the top of the U bracket are holes for mounting a servomotor horn, which is used to connect the steering servomotor.

Figure 8.16 is a diagram of the base with a cutout for the 42-oz servomotor. The base measures 3 in × 5.5 in. The base will hold the power supply and the electronics. Follow the servomotor diagram in removing metal from the base.

First drill the four ($\frac{1}{8}$-in) holes for mounting the servomotor. Next use the same drill bit to drill holes along the inside perimeter of the servomotor cutout. Removing metal in this way is a little easier than trying to saw or nibble it away. When you have drilled as many holes as possible, use the metal nibbler to cut the material between the holes to finish removing this material. Then continue to nibble away at the sides of the cutout until you have the rectangle shape needed. Before you mount the servomotor, file the edges of the hole smooth.

Finish the base by drilling the other holes outlined in the drawing.

The rear axle bracket is shown in Fig. 8.17; it is made from $\frac{1}{8}$- × $\frac{1}{2}$- × 10-in aluminum bar. Drill the four $\frac{1}{8}$-in holes in the aluminum before bending it into shape. For the rear axle I used the wire from a metal coat hanger. Mount the rear axle and wheels to the robot base, using two 6-32 machine screws and nuts.

To continue, we need to mount the front drive wheel to the servomotor. The drive wheel has a diameter of $2\frac{3}{4}$ in and is $\frac{1}{8}$ in thick (see Fig. 8.18). The holes are drilled in the wheel to accept a standard HiTec servomotor horn (see Fig. 8.19). The horn is secured to the wheel using four no. 2 × $\frac{1}{4}$-in sheet metal screws (see Fig. 8.20).

Before you attach the servomotor to the U bracket, secure a servomotor horn to the top of the U bracket, using the predrilled holes (see Fig. 8.21).

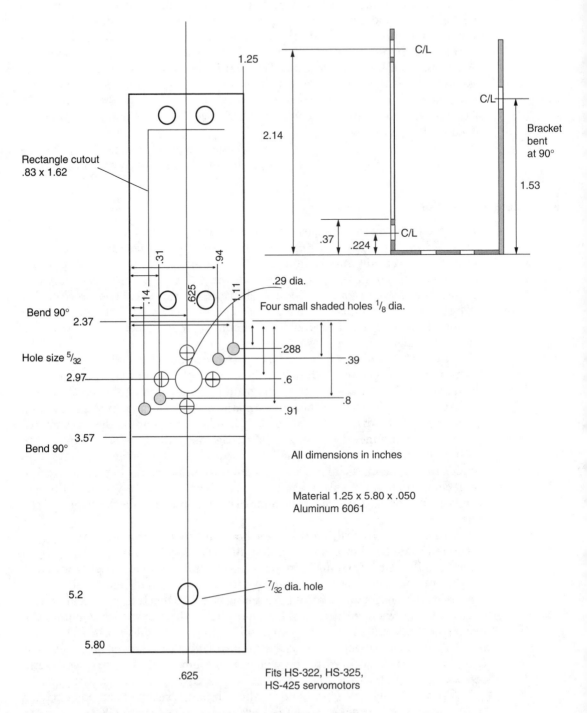

1.25

2.14

C/L

C/L

Bracket
bent
at 90°

1.53

Rectangle cutout
.83 x 1.62

.31

.94

.625

.14

1.11

.29 dia.

Four small shaded holes ¹/₈ dia.

.37

.224

C/L

Bend 90°
2.37

Hole size ⁵/₃₂

2.97

.288

.39

.6

.8

.91

3.57

Bend 90°

All dimensions in inches

Material 1.25 x 5.80 x .050
Aluminum 6061

5.2

⁷/₃₂ dia. hole

5.80

.625

Fits HS-322, HS-325,
HS-425 servomotors

Figure 8.14 Drawing of U bracket for mounting the drive servomotor.

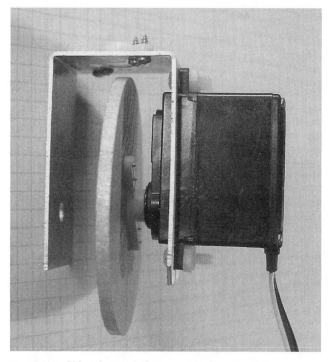

Figure 8.15 U bracket with drive servomotor attached.

The front of the mounting ears, both top and bottom, on the servomotor has small tabs (see Fig. 8.22). Cut and file away these tabs so that the servomotor can be mounted flush against the bracket (see Fig. 8.23). Next mount the servomotor to the U bracket, using 6-32 machine screws and nuts. Attach the wheel/horn assembly to the servomotor (see Figs. 8.24 and 8.25). Put this assembly to the side while we work on other components.

Shell

The original tortoises used a transparent plastic shell. The shell was connected to a bump switch that caused the robot to go into "avoid" mode when activated. I looked at, tried, and rejected a number of different shells. Finally I was left with no choice other than to fabricate my own shell.

Rather than fabricate an entire shell, I made a bumper that encompasses the robot. The bumper is fabricated from $\frac{1}{8}$- $\times$ $\frac{1}{2}$- $\times$ 32-in aluminum bar (see Fig. 8.26). The aluminum bar is marked at the center. Each bend required in the bumper is also marked in pencil. The material is placed in a vise at each pencil mark and bent to the angle required. The two ends of the aluminum bar end up at the center back of the bumper. These two ends are joined together using a $\frac{1}{8}$- $\times$ $\frac{1}{2}$- $\times$ 1-in-long piece of aluminum bar. A $\frac{1}{8}$-in hole is drilled on each end of the aluminum bar. Matching holes are drilled in the ends of the

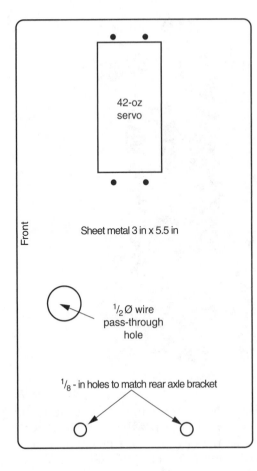

42-oz
servo

Sheet metal 3 in x 5.5 in

Front

$\frac{1}{2}$ Ø wire
pass-through
hole

$\frac{1}{8}$ - in holes to match rear axle bracket

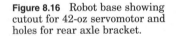

Figure 8.16   Robot base showing cutout for 42-oz servomotor and holes for rear axle bracket.

bumper. The bar is secured to the bumper using two 5-40 machine screws and nuts (see Fig. 8.27).

The upper bracket used to connect the bumper to the robot is identical to the front end of the bumper (see Fig. 8.28). The upper bracket is made from $\frac{1}{8}$- × $\frac{1}{2}$- × 14.5-in aluminum bar. As with the bumper, the center of the bar is marked, and each bend required is also marked in pencil. The material is bent in a vise the same way as the bumper.

Finding the Center of Gravity

It is important to find the center of gravity line of the bumper, because this will mark the optimum location where the upper bracket should be attached. Rest the bumper on a length of aluminum bar. Move the bumper back and forth until it balances evenly on the aluminum bar. Mark the centerline positions on each side of the bumper. Drill a $\frac{1}{8}$-in hole on each side. Drill match-

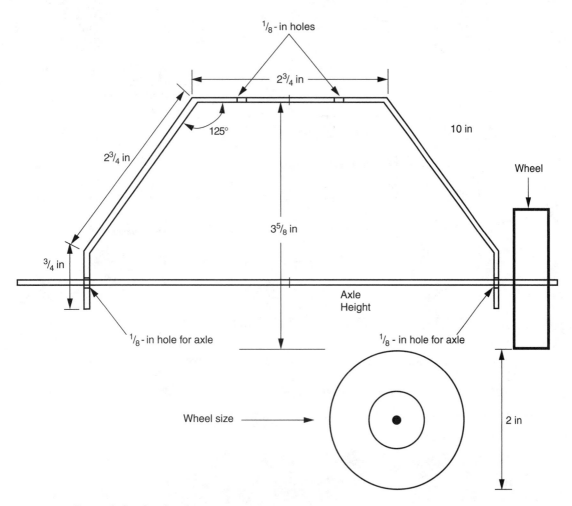

$^1/_8$ - in holes

$2^3/_4$ in

125°

10 in

Wheel

$2^3/_4$ in

$3^5/_8$ in

$^3/_4$ in

Axle
Height

$^1/_8$ - in hole for axle

$^1/_8$ - in hole for axle

Wheel size

2 in

Figure 8.17   Rear axle bracket detail.

ing holes on the ends of the upper bracket. Then secure the upper bracket to the bumper using 5-40 machine screws and nuts.

Attaching Bumper to Robot Base

The bumper is attached to the robot body by the upper bracket. Drill three $^1/_8$-in holes in the top of the upper bracket. One $^1/_8$-in hole is in the center, and the two other holes are $1^1/_8$ in away from the center hole (see Fig. 8.29). Three matching holes are drilled in the robot base behind the servomotor. The holes should be placed so that the bumper (once secured to the base) has adequate clearance ($^1/_8$ to $^1/_4$ in) from the back wheels. The matching center hole on the base must be offset by moving the drilled hole forward on the base by about $^1/_4$ in.

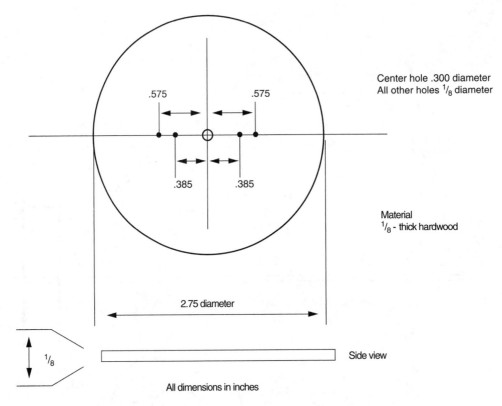

Center hole .300 diameter
All other holes $\frac{1}{8}$ diameter

.575   .575

.385   .385

Material
$\frac{1}{8}$ - thick hardwood

2.75 diameter

$\frac{1}{8}$

Side view

All dimensions in inches

Figure 8.18   Drawing of drive wheel.

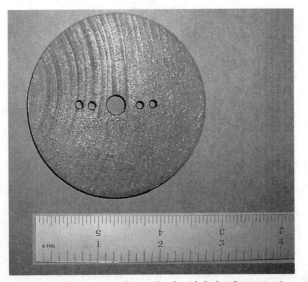

Figure 8.19   Servomotor drive wheel with holes for mounting servomotor horn.

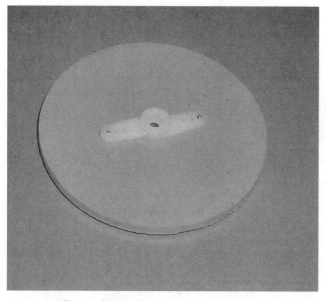

Figure 8.20 Drive wheel with servomotor horn attached.

Figure 8.21 U bracket before mounting of drive servomotor.

The bracket is secured to the base using two 1-in-long 6-32 machine screws, four 6-32 nuts, and two 1-in-long 2-lb compression springs, with a $\frac{1}{8}$-in center diameter (see Fig. 8.30). The tension and resiliency of the bumper can be adjusted by tightening or releasing the upper 6-32 machine screw nuts. Once assembled, the bumper will tilt back and close the bumper switch when the robot (bumper) encounters (pushes against) an obstacle.

Figure 8.22   Tab on servomotor case that needs to be filed off.

Figure 8.23   Tab files off servomotor case.

Bumper Switch

The bumper switch makes use of the center holes. Looking back at Fig. 8.30, we see the center hole is fitted with a 6-32 machine screw held on by a standard (zinc-plated) nut, followed by a brass nut. The brass nut has a wire soldered to it. The purpose of this little assembly is just to attach a wire to the bracket-bumper assembly. Brass nuts are used because it is possible to solder wires to brass to make electrical connections. This is in contrast to the standard zinc-plated steel nuts that are very difficult (impossible) to solder.

Figure 8.24 Attaching drive servomotor to U bracket by using plastic screws and nuts.

Figure 8.25 Another view of drive servomotor and U bracket.

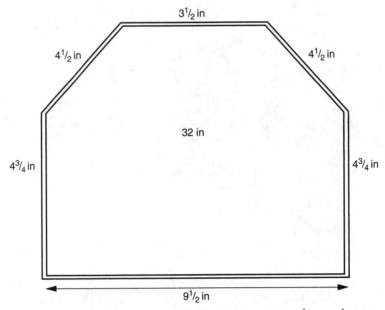

Figure 8.26   Top dimensional view of bumper fabricated from $\frac{1}{8}$-in $\times$ $\frac{1}{2}$-in $\times$ 32-in aluminum bar.

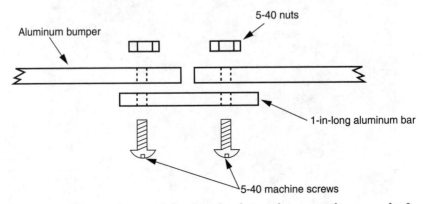

Figure 8.27   Cutaway close-up of aluminum bracket used to secure the open ends of the bumpers.

The second half of the tile switch is comprised of a 1-in 6-32 plastic machine screw and three 6-32 machine screw nuts. One nut must be brass with a wire soldered to it (see Fig. 8.31). Figures 8.32 and 8.33 are close-up photographs of the finished bumper switch. The assembly is adjusted so that the brass nut on the top of the 6-32 machine screw lies just underneath the upper aluminum bracket without touching. When the upper bracket tilts forward, contact is made between the aluminum bracket and brass nut, which is read as a switch closure.

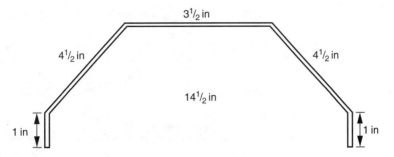

Figure 8.28 Side dimensional view of upper bracket fabricated from $\frac{1}{8}$-in $\times$ $\frac{1}{2}$-in $\times 14\frac{1}{2}$-in aluminum bar.

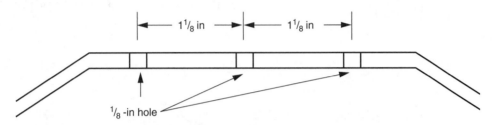

Figure 8.29 Side dimensional view for hole placement in top of the upper bracket.

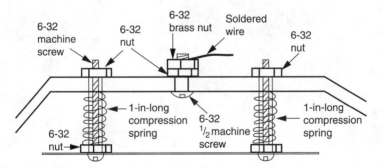

Figure 8.30 Side view of upper bracket detailing the mounting of the upper bracket to the robot base using machine screws and compression springs. Also details bracket half of the bumper switch.

Mounting the Steering Servomotor

If you haven't done so, mount the steering servomotor to the robot base, using four 6-32 plastic machine screws and nuts. Before you attach the U bracket to the steering servomotor, make sure the steering servomotor spindle is in its center position. This will ensure that the robot will steer forward right and left

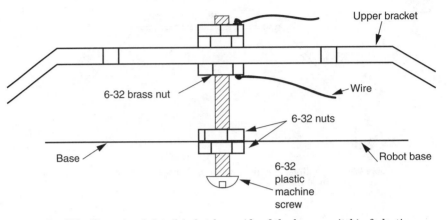

Figure 8.31 Side dimensional detail (robot base side of the bump switch) of plastic screw with top brass nut.

Figure 8.32 Close-up photograph detailing bump switch and spring mounting of upper bracket.

properly. The following short program will place a servomotor in its center position:

```
start:
pulsout portb.1, 150
pause 18
goto start
```

The output pulse signal for the servomotor is taken as pin RB1. Once the servomotor is in its center position, attach the U bracket to the servomotor so that the drive wheel is pointing forward.

Figure 8.33 Close-up photograph detailing bump switch.

Photoresistor

The CdS photoresistors (see Fig. 8.34) used in this robot have a dark resistance of about 100 kΩ and a light resistance of 10 kΩ. The CdS photoresistors have large variances in resistance between cells. It is useful to use a pair of CdS cells for this robot that matches, as best as one can match them, in resistance.

Since the resistance value of the CdS cells can vary so greatly, it's a good idea to buy a few more than you need and measure the resistances, to find a pair whose resistances are close. There are a few ways you can measure the resistance. The simplest method to use a volt-ohmmeter, set to ohms. Keep the light intensity the same as you measure the resistance. Choose two CdS cells that are closely matched within the group of CdS cells you have.

The second method involves building a simple PIC16F84 circuit connected to an LCD display. The advantage of this circuit is that you can see the response of the CdS cells under varying light conditions. In addition, you can see the difference in resistance between the CdS cells when they are held under the same illumination. This numeric difference of the CdS cells under exact lighting is used as a fudge factor in the final turtle program. If you just test the CdS cells with just an ohmmeter, you will end up using a larger fudge factor for the robot to operate properly.

The schematic for testing the CdS cells is shown in Fig. 8.35. The circuit, built on a PIC Experimenter's Board, is shown in Fig. 8.36. The PicBasic Pro testing program follows:

```
'CdS cell test
'PicBasic Pro program
'Serial communication 1200 baud true
'Serial information sent out on port b line 0
```

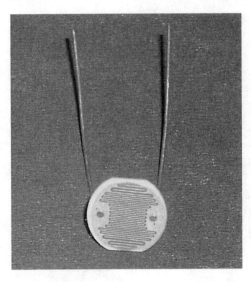

Figure 8.34 CdS photoresistor cell.

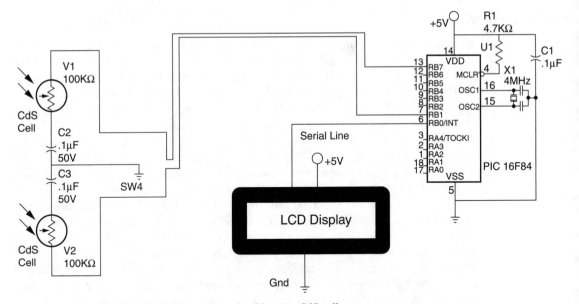

Figure 8.35 Electrical schematic for testing and calibrating CdS cells.

```
'Read CdS cell #1 on port b line 1
'Read CdS cell #2 on port b line 7
v1 var byte                    'Variable v1 holds CdS #1 information
v2 var byte                    'Variable v2 holds CdS #2
```

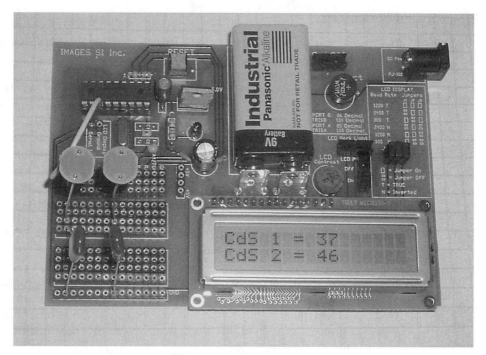

Figure 8.36 Test circuit built on PIC Experimenter's Board.

```
information
pause 1000                              'Allow time for LCD display
main:
pot portb.1,255,v1                      'Read resistance of CdS #1 photocell
pot portb.7,255,v2                      'Read resistance of CdS #2 photocell
'Display information
serout portb.0,1,[$fe,$01]              'Clear the screen
pause 25
serout portb.0,1,["CdS 1 = "]
serout portb.0,1,[#v1]
serout portb.0,1,[$fe,$C0]              'Move to line 2
pause 5
serout portb.0,1,["CdS 2 = "]
serout portb.0,1,[#v2]
pause 100
goto main
```

Notice in Fig. 8.36 that CdS cell 1 is reading 37 and CdS cell 2 is reading 46 under identical lighting. Keep in mind that this is a closely matched pair of CdS cells. We can use a fudge factor of ±15 points. This means that as long as the readings between cells vary from each other by ±15 points, the microcontroller will consider them numerically equal.

Trimming the Sensor Array

If you are using the Experimenter's Board, you can trim and match the CdS cells to one another. Doing so allows you to reduce the fudge factor and produces a crisper response from the robot.

Typically one CdS cell resistance will be lower than that of the other CdS cell. To the lower-resistance CdS cell add a 1-kΩ (or 4.7-kΩ) trimmer potentiometer in series (see Fig. 8.37). Adjust the potentiometer (trim) resistance until the outputs shown on the LCD display equal each other. Trim the CdS cell under the same lighting conditions in which the robot will function. The reason for this is that when the light intensity varies from that nominal point to which you've trimmed the CdS cell, the responses of the individual CdS cells to changes in light intensity also vary from one another and then are not as closely matched.

Once you have a pair of CdS cells to use, they need to be attached to the robot. I soldered the CdS cells and capacitors to a small piece of perforated board (see Fig. 8.38). Figure 8.38 shows both the front and back of the sensor array.

The opposite side of the servomotor bracket that holds the continuous rotation servomotor is perfect for mounting the photoresistor. I used a small piece of transparent plastic, $\frac{1}{2}$ in wide $\times$ 6 in long $\times$ $\frac{1}{16}$ in thick (12.5 mm $\times$ 152 mm $\times$ 1.5 mm thick) to create an L bracket on which to mount the photoresistors (see Fig. 8.39).

A $\frac{1}{8}$-in hole is drilled $\frac{1}{2}$ in up from one end (see Fig. 8.37). The plastic is then gently heated about $2\frac{1}{2}$ in up from the end (see bend point). When the plastic softens, bend it to a 90° angle and hold it in position until the plastic hardens again.

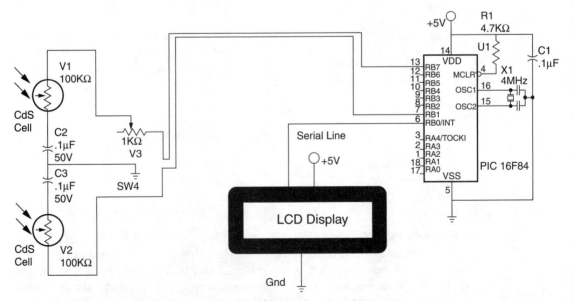

Figure 8.37 Electrical schematic of testing circuit with potentiometer trimmer.

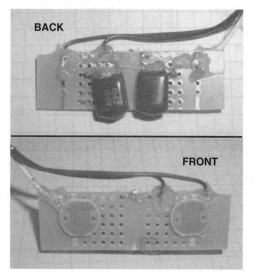

Figure 8.38 Front and back mounting of CdS cells and capacitors to perforated board.

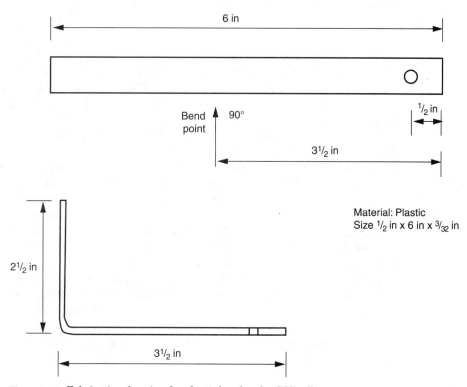

Material: Plastic
Size $1/2$ in x 6 in x $3/32$ in

Figure 8.39 Fabrication drawing for plastic bracket for CdS cells.

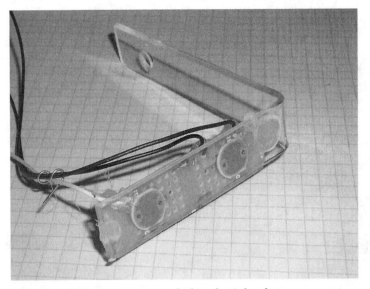

Figure 8.40 CdS sensor array attached to plastic bracket.

Next I used hot glue to secure the CdS cells to the back of the plastic L (see Fig. 8.40). Then I mounted an opaque vane on the front surface of the plastic in between the photoresistors (see Fig. 8.41). The opaque vane is made from a small piece of conductive foam I had lying around. I simply hot-glue one edge to the plastic.

Using the opaque vane and the two CdS photosensors in this configuration alleviates much of the computation needed to track a light source. The operation of the sensor array is shown in Fig. 8.42. When both sensors are equally illuminated, their respective resistances are approximately the same. As long as each sensor is within ±10 points of the other, the PIC program will see them as equal and won't move the servomotor (steering). When the sensor array is not properly aimed at the light source, the vane's shadow falls on one of the CdS cells. This pushes the resistance beyond the ±10-point range. The PIC microcontroller activates the steering servomotor to bring both sensors back under even illumination. In doing so, this steers the robot straight to the light source.

If the sensors detect too great a light intensity, the robot will go into avoid mode.

Mounting the photoresistor array on the drive wheel assembly keeps the sensors pointing in the same direction as the drive wheel (see Fig. 8.43). This replicates the function of the original tortoise robots. The array is secured to the U bracket by using a small plastic screw and wing nut.

Schematic

The schematic for the robot is shown in Fig. 8.44. Intelligence for the robot is provided by a single PIC 16F84 microcontroller. The forward servomotor is

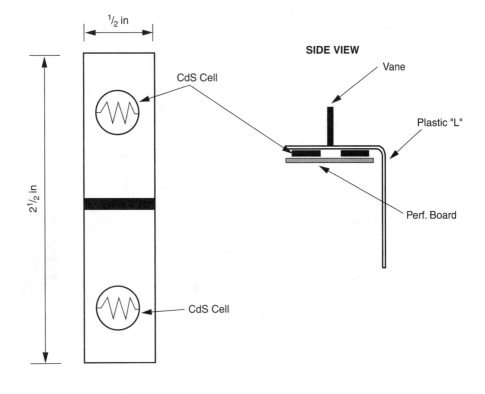

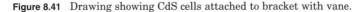

TOP VIEW

Figure 8.41   Drawing showing CdS cells attached to bracket with vane.

connected to RB7, and the steering servomotor control signal is provided by RB6. Sensor readings of the CdS cell are read off pins RB2 and RB3. The bumper switch is read off pin RA0.

There is nothing critical about the circuit; it may be hardwired on a prototyping board. I chose a simpler route. Images SI Inc. sells a four-servomotor controller board. This board has all the connections needed for the sensors and servomotors. My connections to the PC board are shown in Fig. 8.45. A picture of the finished circuit is shown in Fig. 8.46. Notice in the picture I used terminal blocks to connect the sensor array and bumper switch.

Program

Upon power up, the drive motor is off, and the microcontroller begins scanning for the brightest light source, using the servomotor.

If a light source is too bright, the robot jumps into avoid mode. In avoid mode the robot backs away from the light source by reversing the drive motor while

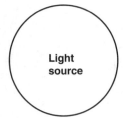

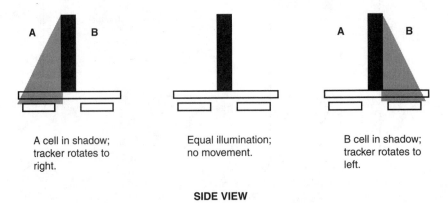

A cell in shadow; tracker rotates to right.

Equal illumination; no movement.

B cell in shadow; tracker rotates to left.

SIDE VIEW

Figure 8.42 Operation of sensor array for targeting light source.

steering the drive wheel left or right. If the light isn't so bright as to activate the avoid mode, the robot steers in the direction of the light and activates the drive wheel forward.

If the bumper switch is activated, the robot assumes it has hit an obstacle and so goes into avoid mode. The robot uses avoid mode for too bright a light and collisions. If the tilt switch is not activated (no collision), then the program jumps to the beginning and the process continues scanning and moving to the brightest light source.

The program is written for the PicBasic Pro compiler that is programmed into a PIC 16F84. The program should be able to be compiled and run with few modifications on the PicBasic version. In-group variances in CdS sensors, drive motors, robot structure, and the like can be adjusted for or modified in the program.

```
'Turtle program
'PicBasic Pro program
'Read CdS cell #1 on port b line 1
'Read CdS cell #2 on port b line 7
v1 var byte              'Variable v1 holds CdS #1 information
```

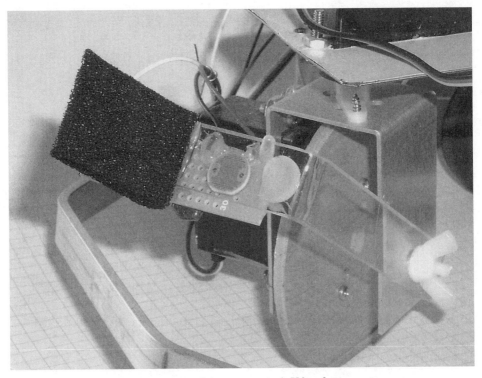

Figure 8.43 Attaching sensory array to drive servomotor's U bracket.

```
v2 var byte              'Variable v2 holds CdS #2 information
v3 var byte              'Variable for calculation
s1 var byte              'Variable s1 holds servomotor #1 pulse width info
s2 var word              'Variable for random function
rv var byte              'Variable rv holds the range value
s1 = 150                 'Initialize steering servomotor facing forward
rv = 10                  'Adjust as needed for smooth operation
ct var byte              'Counter

'Drive servomotor ** continuous rotation information
'Connected to pin portb.7 ** variable pulse width numbers
'157 forward * 165 slow forward
'167 stop
'169 slow backward * 177 backward

start:
pot portb.2,255,v1       'Read resistance of CdS #1 photocell
pot portb.3,255,v2       'Read resistance of CdS #2 photocell

'Check bumper switch "Did I hit something?"
if porta.0 = 0 then avoid              'Hit obstacle go into avoid mode
'Is it sleepy time?
```

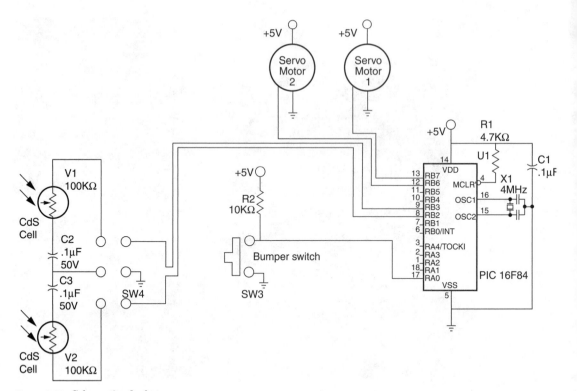

Figure 8.44 Schematic of robot.

```
if v1 <= 230 then skp      'Is it dark enough to sleep?
if v2 > 230 then slp       'Yes
'Is it too bright to see?
skp:                       'No sleep--keep moving
if v1 >= 12 then skip2      'Is it too bright to live?
if v2 < 12 then avoid       'Yes

'Which way do I go?
skip2:                     'Not so bright--should I steer?
if v1 = v2 then straight    'Light is equal go straight
if v1 > v2 then greater     'Check light intensity to turn right
if v1 < v2 then lesser      'Check light intensity to turn left

straight:                  'Go forward in the direction you're facing
pulsout portb.6, s1         'Don't move steering
pulsout portb.7, 157        'Go forward
goto start

greater:
v3 = v1 - v2               'Check numerical difference between CdS cells
if v3 > rv then right       'If more than rv turn right
goto straight               'If not go straight
```

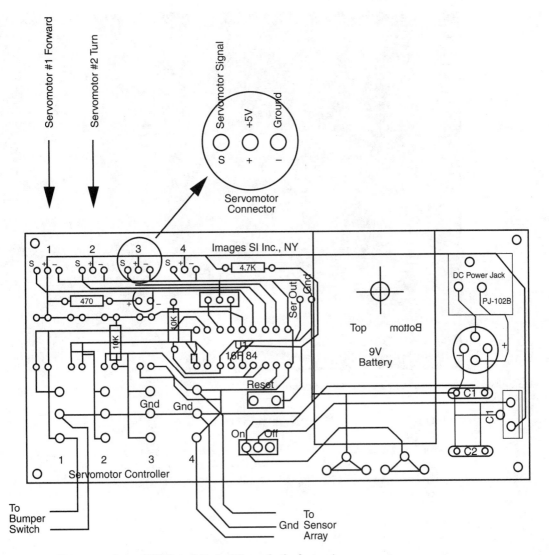

Figure 8.45 Using an existing PCB board for building robot's electronics.

```
lesser:
v3 = v2 - v1                    'Check numerical difference between CdS cells
if v3 > rv then left            'If more than rv turn left
goto straight                   'If not go straight

right:                          'Turn right
s1 = s1 + 1                     'Increment variable s1 to turn right
if s1 > 225 then s1 = 225       'Limit s1 to 225
pulsout portb.6, s1             'Move steering servomotor
pulsout portb.7, 165            'Go forward slowly
goto start
```

Figure 8.46 Close-up of electric circuit board.

```
left:                          'Turn left
s1 = s1 - 1                    'Decrement variable s1 to turn left
if s1 < 65 then s1 = 65        'Limit s1 to 65
pulsout portb.6, s1            'Move steering servomotor
pulsout portb.7, 165           'Go forward slowly
goto start

slp:                           'Go asleep
pulsout portb.6, s1            'Don't move steering
pulsout portb.7, 167           'Stop drive servomotor
goto start

avoid:                         'Avoid mode, send
random s2                      'Randomize s2
s1 = s2 / 256                  'Reduce range of s1 to 1 to 255
if s1 < 65 then s1 = 65        'Set lower limit
if s1 > 225 then s1 = 225      'Set upper limit
for ct = 1 to 125              'Start counter
pulsout portb.6, s1            'Steer (turn) in a random direction
pulsout portb.7, 177           'Reverse drive motor (slow)
pause 18                       'Pause to send instructions at 50 Hz
next ct                        'Loop
```

```
s1 = 150                        'Steer back to center
goto start
```

Adding Sleep Mode

I added a sleep mode for occasions when the ambient light is very low. The robot moves forward when both CdS sensors receive approximately the same light intensity. The robot steers right or left when one CdS cell receives more light than the other. If each CdS cell receives too much light or the bump switch is activated, the robot goes into avoid mode.

Power

A 9-V battery on the PC board supplies adequate electrical power for the robot for a short time. Although I used this power supply for testing robot function, you will need a stronger power supply for extended use. The PCB board has a dc voltage socket where an external power supply can be connected.

The finished robot is shown in Figs. 8.47 and 8.48.

Behavior

This robot exhibits the following behavior. In ambient light, no bright light source, the robot travels in a straight line (or circle depending upon the last light source target). If the ambient light is too bright, it jerks backward. With a mediocre light source, it will aim and travel toward the light.

The program can be developed further to explore more interesting and exotic behaviors. Before we do so, let's first look at how the standard program functions.

Fudge Factor

The variable RV (range value) is the fudge factor. At the beginning of the program the variable RV is assigned a value of 10. In my prototype I actually used an RV of 2 because I had matched the resistance values of CdS cells, as discussed earlier.

Tolerance between the two CdS photoresistors may be increased or decreased by modifying the numerical value of this variable. You may need to adjust this variable according to how closely the resistance values of your CdS cells match.

Light Intensity

The program continually checks the light intensity received (resistance) by each CdS sensor and then makes a decision based on those readings. The maximum reading from the sensor is 255 (total darkness). If the room gets dark enough to generate a value of 230 in each CdS cell, then the robot goes into sleep mode.

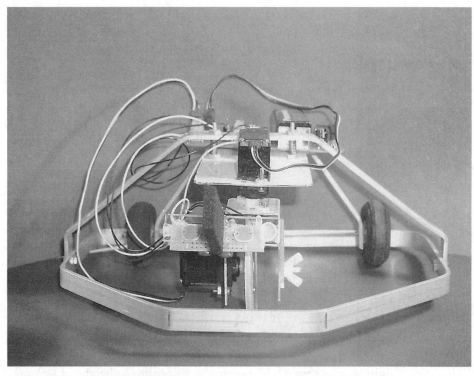

Figure 8.47 Front view of turtle robot.

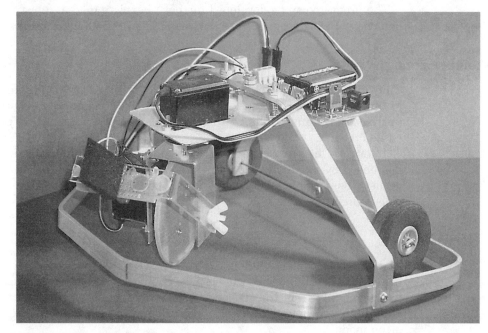

Figure 8.48 Side view of turtle robot.

```
'Is it sleepy time?
if v1 <= 230 then skp      'Is it dark enough to sleep?
if v2 > 230 then slp       'Yes
```

The opposite of sleep mode is the "too bright to live." If the light becomes too intense, this triggers the avoid mode.

```
'Is it too bright to see?
skp:                        'No sleep—keep moving
if v1 >= 12 then skip2      'Is it too bright to live?
if v2 < 12 then avoid       'Yes
```

Increasing the numerical value, in this case 12, decreases the light intensity that puts the robot into avoid mode. Decreasing the numerical value increases the light intensity needed to throw the robot into avoid mode. In most cases you will want to *decrease* this number. However, I would advise you not to go below a numerical value of 9, because even at full light saturation of the CdS cell, its resistance never drops to zero. And in my light saturation tests the sensor never yielded a value less than 5.

In this robot, intense light pushes the robot into avoid mode. If this were a true photovore robot, high light intensity would put it into a feeding mode.

Handedness

In addition, one could create handedness in the robot (right- or left-handed) by modifying either greater or lesser subroutines, not both. This will create a robot that is more likely to turn in one direction than in the other.

```
greater:
v3 = v1 - v2               'Check numerical difference between CdS cells
if v3 > rv then right      'If more than rv turn right
goto straight              'If not go straight
```

For instance, if RV = 10, we can substitute the value 7 like this

```
if v3 > 7 then right
```

Then in the lesser subroutine the RV is kept the same. The result of this manipulation is that we would create a robot that is more likely to turn to the right.

This robot offers opportunities to the robotists and experimenters for continued experimentation and development in both hardware and software.

Parts List

12-in × 12-in sheet metal sheet of 22 or 24 gauge

(1) Aluminum bar $\frac{1}{8}$ in × $\frac{1}{2}$ in × 32 in long

(1) Aluminum bar $\frac{1}{8}$ in × $\frac{1}{2}$ in × $14\frac{1}{2}$ in long

(1) Aluminum bar $\frac{1}{8}$ in $\times$ $\frac{1}{2}$ in $\times$ 2 in long

(1) 42-oz-torque hobby servomotor (HS-322)

(1) Hobby servomotor (HS-425)

(1) $2\frac{3}{4}$-in-diameter drive wheel

(2) CdS photocells, 100 kΩ dark, 10 kΩ light

(1) 10-kΩ, $\frac{1}{4}$-W resistor

(1) 4.7-kΩ, $\frac{1}{4}$-W resistor

(2) 22-pF caps

(1) 4-MHz ceramic resonator or Xtal

(1)(IC1) PIC microcontroller (16F84-04)

(1) U bracket for drive servomotor

Miscellaneous needs include perforated board, $\frac{1}{16}$-in-thick transparent plastic, 5-40 machine screw and nuts, plastic 6-32 $\times$ 1-in machine screw, 6-32 brass nuts, 1-in-long compression springs (2 lb). Aluminum bars, machine screws, tubing, and compression springs are available in most well-stocked hardware stores.

Servomotors may be purchased at hobby shops or electronics distributors.

Electronic components may be purchased from RadioShack, Images SI Inc., Jameco Electronics, JDR Microdevices (see Suppliers at end of book).

PC board, servomotor drive wheel, and U bracket for drive servomotor may be purchased from Images SI Inc.

9

Braitenberg Vehicles

In 1984 Valentino Braitenberg published a book titled *Vehicles—Experiments in Synthetic Psychology*. In his book Valentino describes a number of wondrous vehicles that exhibit interesting behaviors based on the use of a few electronic neurons.

Similar in concept to Walter's seminal neural work with his robot tortoises, Valentino's vehicle behavior is more straightforward, making it somewhat easier to follow both theoretically and logically. This also makes it easier to implement his ideas into real designs for robots.

In this chapter we will build a few Braitenberg-type vehicles.

At the heart of Braitenberg vehicles is his description of a basic vehicle, which is a sensor connected to a motor. Braitenberg continues to explain the relationship between the sensor and motor. The relationship is essentially the connection between the sensor and motor, and this connection ought to be considered as a neuron. With the connection configured as a neuron, the structure is shown in Fig. 9.1. Instead of a vehicle we will describe the structure diagram as a small neural network.

At the front end of the network we find a sensor, followed by the neuron and finally the output motor. The sensor detects the intensity of light and outputs a proportional signal to the motor. High-intensity light produces high rpm's (revolutions per minute) from the motor. Low-intensity light produces slow rpm's.

Consider the sensor portion as modular and interchangeable. Other sensors can be plugged in and incorporated to detect any number of environmental variables, for example, heat, pressure, sound, vibration, magnetic fields (compass), electrical fields, radioactivity, and gases (toxic or otherwise).

In addition, the motor, like the sensor, represents a singular example of an output module. Other output modules could include a second neuron (or neural layer), electric circuit, on/off switch, light source, etc.

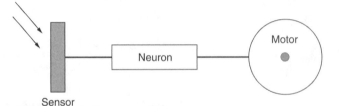

Figure 9.1   Basic neuron setup, sensor input, neuron, and motor output.

The neuron's input is the output of the sensor, and the neuron's output activates a motor in relationship to its input. The input/output "relationship" of the neuron can be made to be one of many different mathematical functions. The relationship may also be called *connection strength* or *connection function* when you are reading the neural network literature. The relationship is one of the most important variables we can modify when programming our robot.

Neural I/O Relationships

When the neuron is stimulated, it generates an output. As stated, there are a number of mathematical functions that can exist inside the neuron. These functions act upon the neuron's input (sensor output) and pass through the results to the neuron's output. Let's examine a few of them.

Positive proportional. As input from the sensor increases, activation (rpm's) of the motor increases in proportion; see Fig. 9.2.

Negative proportional. As input from the sensor increases, activation (rpm's) of the motor decreases in proportion (see Fig. 9.3).

Digital. As input from the sensor output exceeds a predetermined (programmed) threshold (that may be positive or negative), the motor is activated (see Fig. 9.4).

Gaussian. As input from the sensor increases, output passes through a gaussian function for motor activation (see Fig. 9.5).

Essentially the neuron may incorporate any mathematical function. It would perform this function on the sensory input to generate an appropriate output. I have provided an example of only a few of the more common functions available.

Vehicles

Using the basic neural setup, we can construct a few simple vehicles that exhibit interesting behaviors. Figure 9.6 illustrates two vehicles labeled A and B. Both vehicles use the positive proportional neural setup with a light intensity sensor.

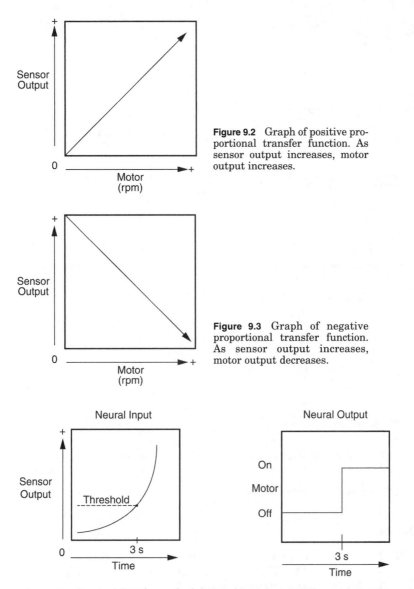

Figure 9.2 Graph of positive proportional transfer function. As sensor output increases, motor output increases.

Figure 9.3 Graph of negative proportional transfer function. As sensor output increases, motor output decreases.

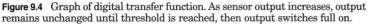

Figure 9.4 Graph of digital transfer function. As sensor output increases, output remains unchanged until threshold is reached, then output switches full on.

Vehicle A, if both sensors are evenly illuminated by a light source, will speed up and, if possible, run into the light source. However, if the light source is off to one side, the sensor on the side of the light source will speed a little faster than the sensor/motor on other side. This will cause the vehicle to veer away from the light source (see Fig. 9.7).

Vehicle B, if both sensors are evenly illuminated by a light source, will speed up and, if possible, run into the light source (same as vehicle A). If the light source is off to one side, vehicle B will turn toward the light source (see Fig. 9.7).

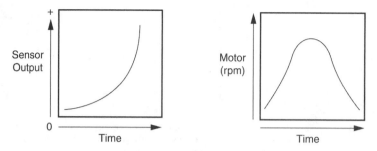

Figure 9.5 Graph of gaussian function. As sensor output increases, output follows a gaussian curve.

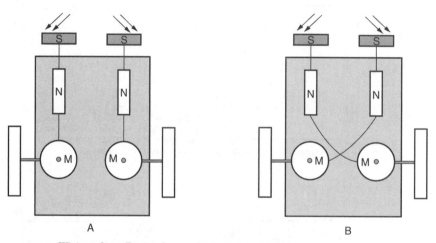

Figure 9.6 Wiring of two Braitenberg vehicles labeled A and B.

Negative proportional neural setups would show the opposite behavior.

Building Vehicles

It's time to put the theory to the test and see if it works. Let's assemble the materials needed to build a vehicle. The photovore's basic operating procedure is like Walter's robot. It tracks and follows a light source.

The base of the vehicle is a sheet of aluminum 8 in long by 4 in wide by $\frac{1}{8}$ in thick. We will use two gearbox motors for propulsion and steering and one multidirectional front wheel.

We will try a new construction method with this robot. Instead of securing the gearbox motors with machine screws and nuts, we will use 3M's industrial brand double-sided tape. This double-sided tape, once cured, is as strong as pop rivets. I tried to separate a sample provided by 3M. It consisted of two flat pieces of metal secured with the tape. Even when I used pliers, it was impossible. 3M states that the tape requires 24 h to reach full strength. You may not achieve the full-strength capability of the tape unless you follow the 3M procedure.

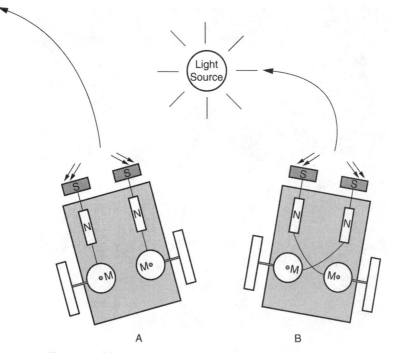

Figure 9.7   Function of A and B Braitenberg vehicles.

The gearbox motor is a 918D type (see Fig. 9.8). The gearbox motor at the top of the picture has an orange cowl that is covering the gears. Notice the flat mounting bracket that is perfect for securing to the vehicle base. The double-sided tape is cut lengthwise to fit the base of bracket to the gearbox motor. The exposed side of the tape is immediately secured to the gearbox motor bracket. Then the motor is positioned on the bottom of the vehicle base, the protective covering of the tape is removed, and the gearbox motor is firmly placed onto the bottom of the vehicle base (see Fig. 9.9).

The second gearbox motor is secured to the other side in a similar manner.

Back wheels

The shaft diameter of the gearbox motor is a little too small to make a good friction fit to the rubber wheel. To beef up the diameter, cut a small 1- to 1.5-in length of the 3-mm tubing; see Parts List. Place the tubing over the gearbox motor shaft, and collapse the tubing onto the shaft, using pliers. There is a small cutaway on the gearbox motor shaft (see Fig. 9.10). If you can collapse the tubing into this cutaway, you will create a strong fit between the shaft and the tubing that will not pull off easily (see Fig. 9.11).

The tubing adds to the diameter of the shaft and will make a good friction fit with the rubber wheels (see Fig. 9.12). Simply push the center holes of the wheels onto the tubing/shaft, and you are finished.

Figure 9.8 A 918D 100:1 Gearbox motor.

Figure 9.9 3M double-sided tape is used to secure gearbox motor to base of vehicle.

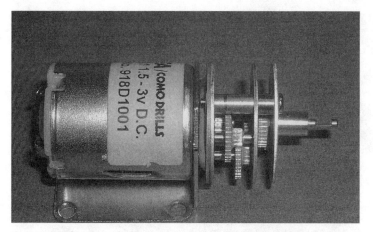

Figure 9.10 Gearbox motor showing cutaway on output shaft.

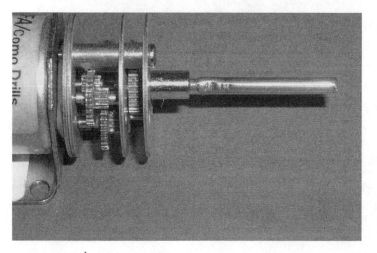

Figure 9.11   A $1\frac{1}{2}$-in length of 3-mm-diameter tubing attached to gearbox motor shaft.

Front wheels

Steering is accomplished by turning on or off the gearbox motors. For instance, turning on the right while the left gearbox motor is off will turn the vehicle to the left, and vice versa. In similar vehicles many times the robotists will forgo front wheels entirely and use a skid instead. This allows the vehicle to turn without concern about the front wheels pivoting and turning in the proper direction

The multidirectional wheel accomplishes much the same thing as a skid, but does so with less resistance. Figure 9.13 shows the multidirectional wheel. It is constructed using rollers around its circumference that allow the wheel to rotate forward and move sideways without turning.

Figure 9.12   Rubber wheel used to friction fit onto gearbox motor shaft.

The multidirectional wheel is attached using a basic U-shaped bracket (see Fig. 9.14). The bracket is secured to the front of the vehicle base using the 3M double-sided tape. The multidirectional wheel is secured inside the U bracket using a small 2.25-in piece of $1/4$-20 threaded rod and two machine screw nuts (see Fig. 9.15).

With the motors and the multidirectional wheel mounted, we are ready for the electronics. Figure 9.16 shows the underside of the Braitenberg vehicle at this point. I drilled a $1/4$-in hole in the aluminum plate to allows wires from the gearbox motors underneath the robot to be brought topside.

The schematic for the electronic circuit is shown in Fig. 9.17. I built the circuit on two small solderless breadboards. You can do the same or hardwire the components to a PC board. The circuit is pretty straightforward. The gearbox motors require a power supply of 1.5 to 3.0 V. Rather than place another voltage regulator into the circuit, I wired three silicon diodes in series off the 5-V dc power. The voltage drop across each diode is approximately 0.7 V. Across the three series diodes ($0.7 \times 3 = 2.1$ V) equals approximately 2.1 V. If we subtract this voltage drop from our regulated 5-V dc power supply, we can supply approximately 3 V dc to the gearbox motors.

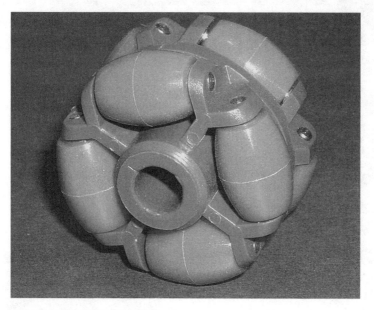

Figure 9.13 Multidirectional wheel.

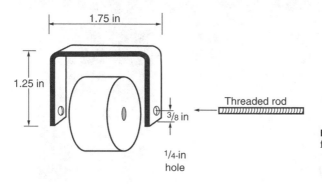

Figure 9.14 Drawing of U bracket for multidirectional wheel.

CdS photoresistor cells

As with Walter's turtle-type robot, we use two CdS photoresistor cells. The CdS photoresistors (see Fig. 9.18) used in this robot have a dark resistance of about 100 kΩ and a light resistance of 10 kΩ. The CdS photoresistors typically have large variances in resistance between cells. It is useful to use a pair of CdS cells for this robot that matches, as best as one can match them, in resistance.

Since the resistance values of the CdS cells can vary so greatly, it's a good idea to buy a few more than you need and measure the resistances to find a pair whose resistances are close. There are a few ways you can measure the resistance. The simplest method to use a volt-ohmmeter, set to ohms. Keep the light intensity the same as you measure the resistance. Choose two CdS cells that are closely matched within the group of CdS cells you have.

Figure 9.15 Multidirectional wheel and U bracket attached to vehicle base.

Figure 9.16 Underside of Braitenberg vehicle showing wheels and gearbox motor drive.

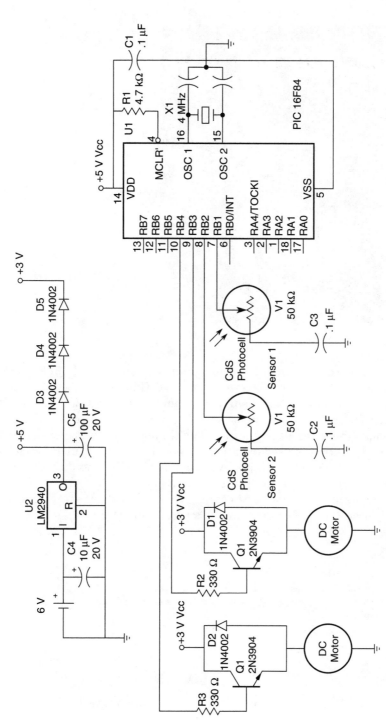

Figure 9.17 Schematic of Braitenberg vehicle.

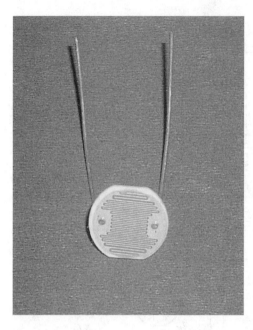

Figure 9.18 CdS photoresistor cell.

The second method involves building a simple PIC 16F84 circuit connected to an LCD display. The advantage of this circuit is that you can see the response of the CdS cells under varying light conditions. In addition, you can see the difference in resistance between the CdS cells when they are held under the same illumination. This numeric difference of the CdS cells under exact lighting is used as a fudge factor in the final turtle program. If you just test the CdS cells with just an ohmmeter, you will end up using a larger fudge factor for the robot to operate properly.

The schematic for testing the CdS cells is shown in Fig. 9.19. The circuit, built on a PIC Experimenter's Board, is shown in Fig. 9.20. The PicBasic Pro testing program follows:

```
'CdS cell test
'PicBasic Pro program
'Serial communication 1200 baud true
'Serial information sent out on port b line 0
'Read CdS cell #1 on port b line 1
'Read CdS cell #2 on port b line 7
v1 var byte                     'Variable v1 holds CdS #1 information
v2 var byte                     'Variable v2 holds CdS #2 information
Pause 1000                      'Allow time for LCD display
main:
pot portb.1,255,v1              'Read resistance of CdS #1 photocell
pot portb.7,255,v2              'Read resistance of CdS #2 photocell
'Display information
serout portb.0,1,[$fe,$01]      'Clear the screen
```

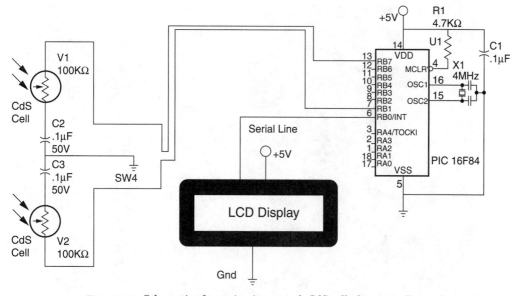

Figure 9.19 Schematic of test circuit to match CdS cells for use in Braitenberg vehicle.

```
pause 25
serout portb.0,1,["CdS 1 = "]
serout portb.0,1,[#v1]
serout portb.0,1,[$fe,$c0]          'Move to line 2
pause 5
serout portb.0,1,["CdS 2 = "]
serout portb.0,1,[#v2]
pause 100
goto main
```

Notice in Fig. 9.20 that CdS cell 1 is reading 37 and CdS cell 2 is reading 46 under identical lighting. Keep in mind, this is a closely matched pair of CdS cells. We can use a fudge factor of ±15 points, meaning that as long as the readings between cells vary from each other by ω15 points, the microcontroller will consider them numerically equal.

Trimming the sensor array

If you are using the Experimenter's Board, you can trim and match the CdS cells to one another. Doing so allows you to reduce the fudge factor and produces a crisper response from the robot.

Typically one CdS cell resistance will be lower than that of the other CdS cell. To the lower-resistance CdS cell add a 1-kΩ (or 4.7-kΩ) trimmer potentiometer in series (see Fig. 9.21). Adjust the potentiometer (trim) resistance until the outputs shown on the LCD display equal each other. Trim the CdS cell under the same lighting conditions in which the robot will function. The

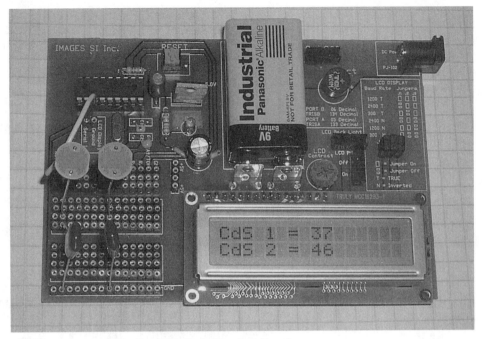

Figure 9.20 Test circuit built on PIC Experimenter's Board.

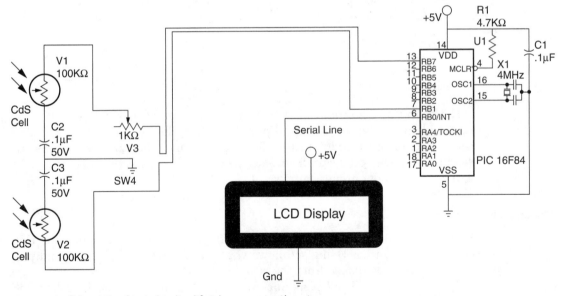

Figure 9.21 Schematic of test circuit with trimmer potentiometer.

reason for this is that when the light intensity varies from that nominal point to which you've trimmed the CdS cell, the responses of the individual CdS cells to changes in light intensity also vary from one another and then are not as closely matched.

PIC 16F84 microcontroller

The 16F84 microcontroller used in this robot simulates two neurons. Each neuron's input is connected to a CdS cell. The output of each neuron activates one gearbox motor.

In the program I put in a fudge factor, or range, over which the two CdS cells can deviate from one another in resistance readings and still be considered equal. If the robot doesn't travel straight ahead when the two CdS cells are equally illuminated, you can increase the range until it does.

PicBasic Compiler program

```
'Braitenberg vehicle 1
start:
pot 1, 255,b0            'Read CdS cell # 1
pot 2, 255,b1            'Read CdS cell # 2
If b0 = b1 then straight
if b0 > b1 then left
if b1 > b0 then right

    straight:
    high 3: high 4
    goto start

    left:
    b2 = b0 - b1          'Compare numerical values +/- 15
    if b2 > 15 then left1  'If greater than 15 turn left
    goto straight          'If not go to straight subroutine
      left1:               'Turn left
        high 3: low 4      'Motor control
    goto start

    right:                 'Compare numerical values +/- 15
    b2 = b1 - b0           'If greater then 15 points
    if b2 > 15 then right1  'Turn toward the right
    goto straight          'If not go straight
      right1:              'Turn right
        high 4: lo3        'Motor control
    goto start             'Do again
```

Testing

The finished robot is shown in Fig. 9.22. For power I used 4 AA cell batteries. I pointed one CdS cell to the left and the other to the right (see Fig. 9.23). To

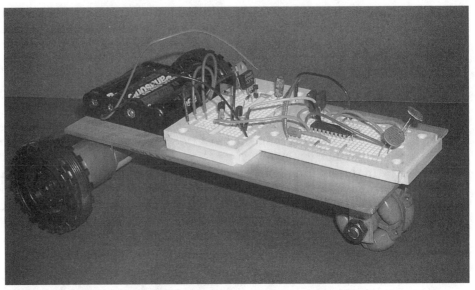

Figure 9.22 Finished Braitenberg vehicle.

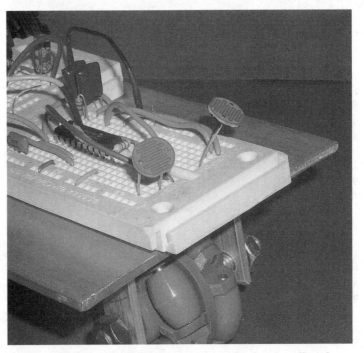

Figure 9.23 Close-up of CdS cells mounted in solderless breadboard.

test the robot's function, I used a flashlight. Using the flashlight, I was able to steer the mobile platform around by shining the flashlight on the CdS cells.

Second Braitenberg Vehicle (Avoidance Behavior)

Given the way the robot is currently wired, it is attracted to and steers toward a bright light source. By reversing the wiring going to the gearboxes you can create the opposite behavior.

Parts List

(1) Microcontroller (16F84)

(1) 4.0-MHz crystal

(2) 22-pF caps

(1) 0.1-μF cap

(1) 100-μF cap

(1) 10-μF cap

(2) 0.1-μF caps

(2) 330-Ω, $\frac{1}{4}$-W resistors

(1) 4.7-kΩ, $\frac{1}{4}$-W resistor

(2) CdS photoresistor cells (see text)

(2) 100:1 gearbox motors (918D)

(2) NPN transistors (2N3904)

(5) Diodes (1N4002)

(2) 2.25-in-diameter wheels

(1) Multidirectional wheel

(1) Voltage regulator (low drop-down voltage +5 V) (LM2940)

Miscellaneous items needed include 6-in length of 3-mm hollow tubing, aluminum 8 in $\times$ 4 in $\times$ $\frac{1}{8}$ in thick, 2 solderless breadboards, 3M double-sided tape, battery holder for 4 D batteries, 3-in $\frac{1}{4}$-20 threaded rod, and 2 machine screw nuts.

10

Hexapod Walker

Legged walkers are a class of robots that imitate the locomotion of animals and insects, using legs. Legged robots have the potential to transverse rough terrains that are impassable by standard wheeled vehicles. It is with this in mind that robotists are developing walker robots.

Imitation of Life

Legged walkers may imitate the locomotion style of insects, crabs, and sometimes humans. Biped walkers are still a little rare, requiring balance and a good deal more engineering science than multilegged robots. A bipedal robot walker is discussed in detail in Chap. 13. In this chapter we will build a six-legged walker robot.

Six Legs—Tripod Gait

Using a six-legged model, we can demonstrate the famous tripod gait used by the majority of legged creatures. In the following drawings a dark circle means the foot is firmly planted on the ground and is supporting the weight of the creature (or robot). A light circle means the foot is not supporting any weight and is movable.

Figure 10.1A shows our walker at rest. All six feet are on the ground. From the resting position our walker decides to move forward. To step forward, it leaves lifts three of its legs (see Fig. 10.1B, white circles), leaving its entire weight distributed on the remaining three legs (dark circles). Notice that the feet supporting the weight (dark circles) are in the shape of a tripod. A tripod is a very stable weight-supporting position. Our walker is unlikely to fall over. The three feet that are not supporting any weight may be lifted (white circles) and moved without disturbing the stability of the walker. These feet move forward.

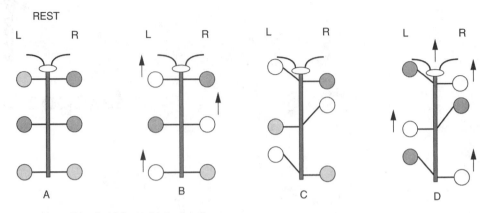

Figure 10.1 Sample biological tripod gait.

Figure 10.1C illustrates where the three lifted legs move. At this point, the walker's weight shifts from the stationary feet to the moved feet (see Fig. 10.1D). Notice that the creature's weight is still supported by a tripod position of feet. Now the other set of legs moves forward and the cycle repeats.

This is called a *tripod gait,* because a tripod positioning of legs always supports the weight of the walker.

Three-Servomotor Walker Robot

The robot we will build is shown in Fig. 10.2. This walker robot is a compromise in design, but allows us to build a six-legged walker using just three servomotors. The three-servomotor hexapod walker demonstrates a true tripod gait. It is not identical to the biological gait we just looked at, but close enough.

This legged hexapod uses three inexpensive HS-322 (42-oz torque) servomotors for motion and one PIC 16F84 microcontroller for brains. The microcontroller stores the program for walking, controls the three servomotors, and reads the two sensor switches in front. The walking program contains subroutines for walking forward and backward, turning right, and turning left. The two switch sensors positioned in the front of the walker inform the microcontroller of any obstacles in the walker's path. Based on the feedback from these switch sensors, the walker will turn or reverse to avoid obstacles placed in its path.

Function

The tripod gait I programmed into this robot isn't the only workable gait. There are other perfectly usable gaits you can develop on your own. Consider

Figure 10.2 Hexapod robot.

this walking program a working start point. To modify the program, it's important to understand both the program and robot leg functions. First let's look at the robot.

At the rear of the walker are two servomotors. One is identified as L for the left side, the other as R for the right side. Each servomotor controls both the front and back legs on its side. The back leg is attached directly to the horn of the servomotor. It is capable of swinging the leg forward and backward. The back leg connects to the front leg through a linkage. The linkage makes the front leg follow the action of the back leg as it swings forward and back.

The third servomotor controls the two center legs of the walker. This servomotor rotates the center legs 20° to 30° clockwise (CW) or counterclockwise (CCW), tilting the robot to one side or the other (left or right).

With this information we can examine how this legged robot will walk.

Moving Forward

We start in the rest position (see Fig. 10.3). As before, each circle represents a foot, and the dark circles show the weight-bearing feet. Notice in the rest position, the center legs do not support any weight. These center legs are made to be $1/_8$ in shorter than the front and back legs.

In position A the center legs are rotated CW by about 25° from center position. This causes the robot to tilt to the right. The weight distribution is now on the front and back right legs and the center left leg. This is the standard tripod position as described earlier. Since there is no weight on the front and back left legs, they are free to move forward as shown in the B position of Fig. 10.3.

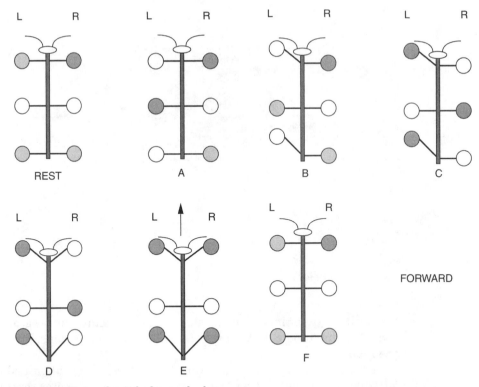

Figure 10.3 Forward gait for hexapod robot.

In the C position the center legs are rotated CCW by about 25° from center position. This causes the robot to tilt to the left. The weight distribution is now on the front and back left legs and the center right leg. Since there is no weight on the front and back right legs, they are free to move forward, as shown in the D position.

In position E the center legs are rotated back to their center position. The robot is not in a tilted position so its weight is distributed on the front and back legs. In the F position, the front and back legs are moved backward simultaneously, causing the robot to move forward. The walking cycle can then repeat.

Moving Backward

We start in the rest position (see Fig. 10.4), as before. In position A the center legs are rotated CW by about 25° from center position. The robot tilts to the right. The weight distribution is now on the front and back right legs and the center left leg. Since there is no weight on the front and back left legs, they are free to move backward, as shown in the B position of Fig. 10.4.

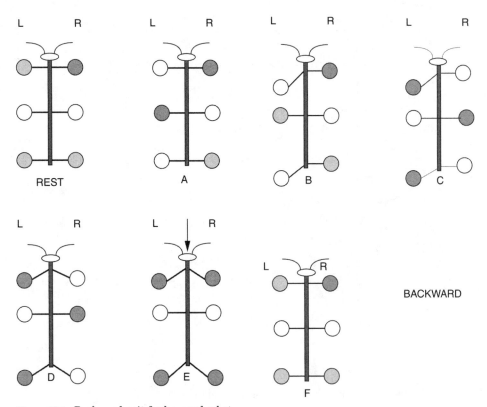

Figure 10.4 Backward gait for hexapod robot.

In the C position the center legs are rotated CCW by about 25° from center position. The robot tilts to the left. Since there is no weight on the front and back right legs, they are free to move backward, as shown in the D position.

In position E the center legs are rotated back to their center position. The robot is not in a tilted position, so its weight is distributed on the front and back legs. In the F position, the front and back legs are moved forward simultaneously, causing the robot to move backward. The walking cycle can then repeat.

Turning Left

The leg motion sequence to turn left is shown in Fig. 10.5. In position A the center legs are rotated CW by about 25° from center position. The robot tilts to the right. The weight distribution is now on the front and back right legs and the center left leg. Since there is no weight on the front and back left legs, they are free to move forward, as shown in Fig. 10.4.

In the B position, the center legs are rotated CCW by about 25° from center position. The robot tilts to the left. Since there is no weight on the front and back right legs, they are free to move backward, as shown in the C position.

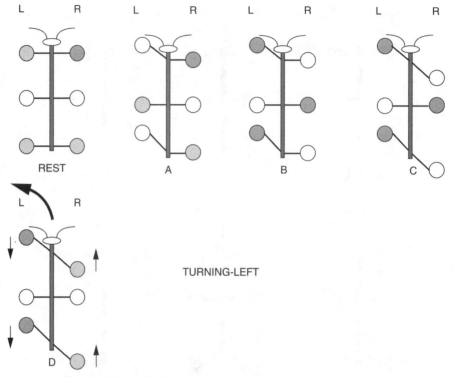

Figure 10.5 Turning-left gait for hexapod robot.

In position D, the center legs are rotated back to their center position. The robot is not in a tilted position, so its weight is distributed on the front and back legs. In position, the left legs moved backward while the right legs moved forward, simultaneously causing the robot to turn left. It typically takes three turning cycles to turn the robot 90°.

Turning Right

Turning right follows the same sequence as turning left, with the leg positions reversed.

Construction

For the main body I used a sheet of aluminum 3 in wide × 9 in long × 0.032 in thick. The servomotors are mounted to the front of the body (see Fig. 10.6).

The four $^{11}/_{64}$-in-diameter holes a little past halfway down the main body are for mounting the center servomotor. These four holes are offset to the right side. This is necessary to align the servomotor's horn in the center of the body.

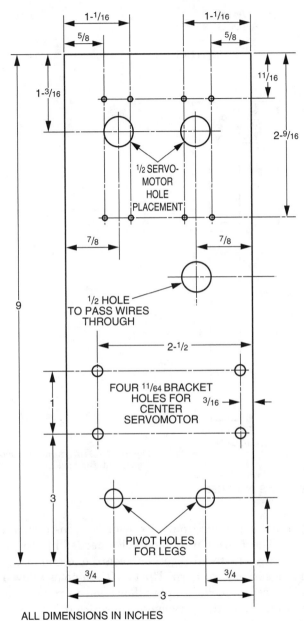

1-1/16   1-1/16

5/8   5/8

11/16

1-3/16

2-9/16

1/2 SERVO-
MOTOR
HOLE
PLACEMENT

7/8   7/8

1/2 HOLE
TO PASS WIRES
THROUGH

2-1/2

9

FOUR 11/64 BRACKET
HOLES FOR
CENTER
SERVOMOTOR

3/16

1

3

1

PIVOT HOLES
FOR LEGS

3/4   3/4

3

ALL DIMENSIONS IN INCHES

Figure 10.6   Diagram of robot base.

The bottom two holes are for mounting the pivots for the two back legs.

Use a punch to dimple the metal in the center of each hole you plan to drill. This will prevent the drill bit from walking when you drill the hole. If you don't have a punch available, use the pointed tip of a nail for a quick substitute.

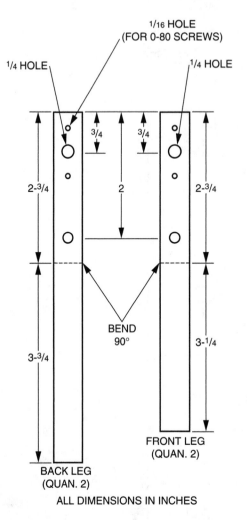

1/16 HOLE
(FOR 0-80 SCREWS)

1/4 HOLE

1/4 HOLE

3/4   3/4

2-3/4   2   2-3/4

BEND
90°

3-1/4

FRONT LEG
(QUAN. 2)

Figure 10.7 Diagram of robot legs (front and back).

3-3/4

BACK LEG
(QUAN. 2)

ALL DIMENSIONS IN INCHES

The legs for the robot are made from $\frac{1}{2}$-in-wide × $\frac{1}{8}$-in-thick aluminum bar stock (see Fig. 10.7). There are four drilled holes needed in the two back legs. The three holes that are clustered together toward one end of the leg are for mounting the leg to a servomotor horn. The two $\frac{1}{16}$-in holes allow a 0-80 screw to pass through. The centered $\frac{1}{4}$-in hole allows you to remove or attach the servomotor screw that holds the servomotor horn (and leg assembly) to the servomotor. Make sure these three holes line up with the holes on the servomotor horn you intend to use.

The front legs only need two holes—one for the pivot and the other for the linkage. Also notice that the front legs are 0.25 in shorter than the back legs. This compensates for the height of the servomotor mounting horn on the back servomotors where the back legs are attached. Shortening the front legs makes the robot platform approximately level.

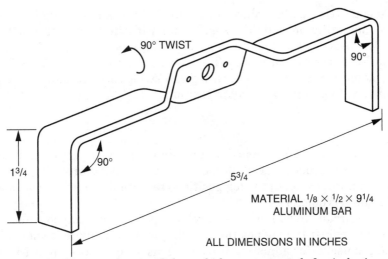

90° TWIST

90°

90°

1³/₄

5³/₄

MATERIAL ¹/₈ × ¹/₂ × 9¹/₄
ALUMINUM BAR

ALL DIMENSIONS IN INCHES

Figure 10.8 Diagram of center tilt legs, which are constructed of a single piece of aluminum and are ¹/₈ in shorter than the front and back legs.

After the holes are drilled, we need to bend the aluminum bar into shape. Secure the aluminum bar in a vise 2³/₄ in from the end with the drilled holes. Pressure is applied to bend the aluminum bar at a 90° angle. It's best to apply pressure at the base of the aluminum bar close to the vise. This will bend the leg at a 90° angle, while keeping the lower portion of the leg straight without any bowing of the lower portion.

The center legs are made from one piece of aluminum (see Fig. 10.8). The center legs are about ¹/₈ in shorter than the front and back legs when mounted to the robot. So when centered, the legs do not support any weight. These legs are for tilting the robot to the left or right. The legs tilt the robot by rotating the center servomotor approximately ±20°.

To produce the center legs, first drill the servomotor horn's mounting holes in the center of the ¹/₈-in × ¹/₂-in × 9¹/₄-in aluminum bar. This should be similar to the three clustered holes you drilled in the back legs. Next secure the aluminum bar in a vise. The top of the vise should hold the aluminum bar ³/₄ in from the center of the aluminum bar. Grab the aluminum bar with pliers about ¹/₂ in above the vise. Keeping a secure grip with the pliers, slowly twist the aluminum bar 90°. Don't go fast, or you could easily snap the aluminum bar. Repeat the twist on the other side.

After the two 90° twists have been made, make the other 90° bend for the legs, as we have done before for the front and back legs.

Mounting the servomotors

The back servomotors are attached to the aluminum body using plastic 6-32 machine screws and nuts. The reason I used plastic screws is that the plastic is a little flexible, allowing the drilled holes to be slightly off-center from the mounting holes on the servomotor without creating a problem.

The legs are attached to the servomotor's plastic horn. For this I used 0-80 machine screws and nuts. When you mount the servomotor horn on the servomotor, make sure that each leg can swing forward and backward an equal amount from a perpendicular position.

Leg positioning

The legs must be positioned accurately, or the walking program will not cause the hexapod robot to walk properly. To aid in this positioning look at Fig. 10.9. The numbers next to the leg positions represent the pulse width output signal for the servomotors.

The circuit we will use to control and power the hexapod walker may also be used to adjust the leg positions. A simplified schematic is shown in Fig. 10.10 that is useful for adjusting the legs. This schematic is almost identical to the schematic that will control the robot; the only difference is that the two sensor switches are removed. The leg adjustment program is small; see below for both PicBasic Pro and PicBasic versions.

If you decide to buy the PCB board for this robot (Fig. 10.22), you can use the PCB board for this test circuit and program.

To align the legs, first disconnect the servomotor horn from the servomotor by unscrewing the center mounting screw from the horn. Once the screw is removed, pull the horn off. Keep the leg attached to the horn. Apply power to the servomotor and connect the control line of the servomotor to RB4. This will center the servomotor's rotational position. Now reattach the servomotor horn to the servomotor, positioning the leg to be in the center position, as shown in Fig. 10.9. Lock the servomotor horn in place, using the center screw. The leg is now in proper position. By con-

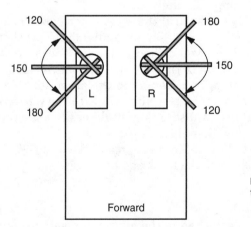

Figure 10.9 Diagram of leg positions relating to pulse widths.

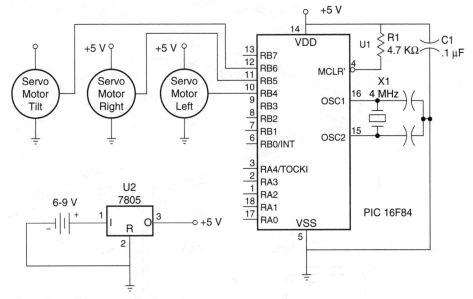

Figure 10.10 Schematic of test circuit.

necting the servomotor control line to pins RB5 and RB6, you can verify the leg's front and back swing. Adjust the program if necessary to ensure a proper swing.

When switching a servomotor from pin to pin, you must power down the circuit first. If you just switch pins without powering down, the microcontroller could latch up and you will get inaccurate positioning.

```
'Leg adjustment program (PicBasic Pro)--for 16f84 microcontroller
start:
pulsout portb.4, 150               'Pin rb4
pulsout portb.5, 120               'Pin rb5
pulsout portb.6, 180               'Pin rb6
pause 18
goto start
end


'Leg adjustment program (PicBasic)--for 16f84 microcontroller
start:
pulsout 4, 150                     'Pin rb4
pulsout 5, 120                     'Pin rb5
pulsout 6, 180                     'Pin rb6
pause 18
goto start
end
```

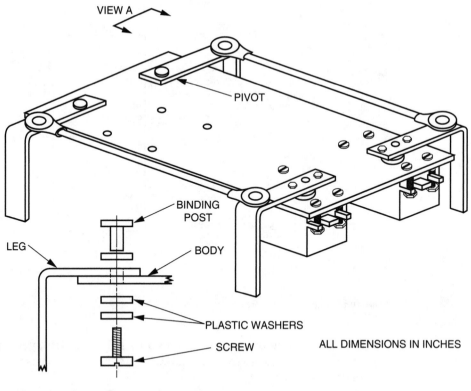

Figure 10.11 Diagram of robot base with front and back leg linkage. View A shows detail of pivot for front legs.

Linkage

The linkage between the front and back legs is made from standard Radio Control (RC) clevis linkage (see Fig. 10.11). In the prototype robot the linkage is $6^3/_4$ in center to center. The linkage fits inside the holes in the front and back legs.

The back legs must be attached to the body of the robot before you make the linkage. The pivot for the front legs is made from a $^3/_8$-in binding post and screw. The leg is attached as shown in the close-up in Fig. 10.11. The plastic washers underneath the body are necessary. They fill up the space between the aluminum body and the bottom of the screw. This keeps the leg close to the aluminum body without sagging. I choose plastic washers for less friction. Do not use so many washers that force is created, binding the leg to the body. The joint should pivot freely.

Center (tilt) servomotor

To attach the center servomotor to the body requires two L-shaped brackets (see Fig. 10.12). Drill the holes and bend at a 90° angle.

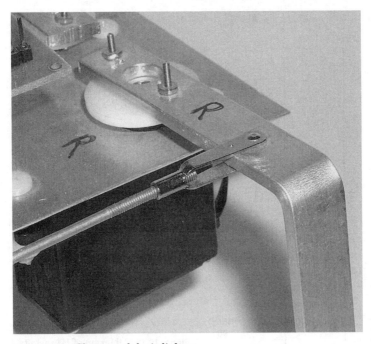

Figure 10.12 Close-up of clevis linkage.

Attach the two L brackets to the center servomotor, using the plastic screws and nuts (see Fig. 10.13). Next mount the center servomotor assembly under the robot body. Align the four holes in the body with the top holes in the L brackets. Secure with plastic screws and nuts.

You must align the center legs on the center servomotor properly, or else the robot will not tilt properly. First remove the horn from the center servomotor. Then attach the center leg to the removed horn, using the 0-80 screws ands nuts. Apply the center control signal (RB4 from Fig. 10.10) to the center servomotor. With the servomotor centered, reattach the horn/center leg assembly to the servomotor, making sure that the legs are in the center position when securing it in position. Once the center leg is attached, you can remove power from the servomotor. Figures 10.14 and 10.15 show the underside and top side of the hexapod robot.

Sensors

This hexapod has two front switch sensors for detecting obstacles (see Fig. 10.16). The switch is a miniature snap-action flat lever arm, model number TFCGV3VT185BC manufactured by C&K Components. The levers on the switches are retrofitted with feelers that extend the range of the levers forward and to the side. The feelers are made with miniature metal tubing or stiff wire (aluminum, steel, or copper).

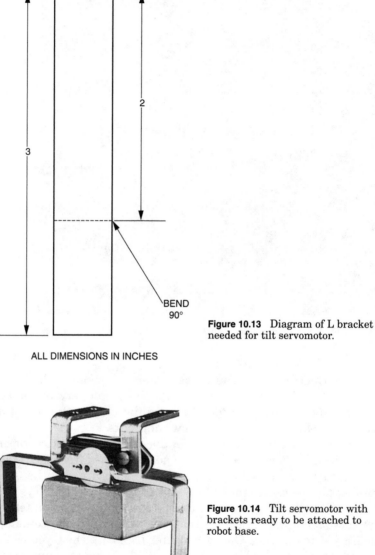

3

2

BEND
90°

ALL DIMENSIONS IN INCHES

Figure 10.13 Diagram of L bracket
needed for tilt servomotor.

Figure 10.14 Tilt servomotor with
brackets ready to be attached to
robot base.

To attach the feelers to the lever, I used a $\frac{3}{8}$-in-long piece of small rubber
tubing. I slid two sections of tubing onto the lever, then slid the stiff wire
underneath the tubing (see Fig. 10.17).

Attaching the switches to the front of the hexapod required a small fixture
to prevent the mounting screws for the switches from getting in the way of the
moving front legs. The fixture is made from two pieces of wood. One piece of
wood measures $\frac{1}{2}$ in wide $\times$ $\frac{1}{4}$ in thick $\times$ 1 in long. The second piece of wood
measures $\frac{3}{4}$ in wide $\times$ $\frac{1}{4}$ in thick $\times$ 3 in long.

Figure 10.15 Tilt servomotor attached to robot base.

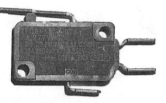

Figure 10.16 Snap-action lever switch used for front obstacle sensors.

Figure 10.17 Bottom view of switch assembly showing feelers.

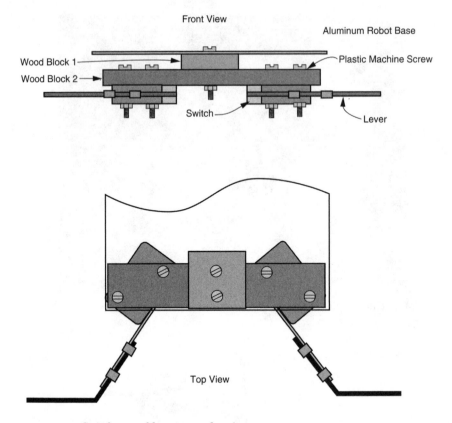

Figure 10.18 Switch assembly cutaway drawing.

Figure 10.18 illustrates the construction of the switch assembly. The two switches are mounted diagonally on the 3-in-long piece of wood using plastic machine screws and nuts. The 1-in-long piece of wood is mounted on top of the 3-in-long piece of wood. Two holes are drilled through the robotic base and two pieces of wood. The assembly is mounted to the robotic base using two plastic machine screws and nuts.

Figures 10.19 and 10.20 show the front and bottom views of the switch assembly.

Electronics

Figure 10.21 shows the schematic for the servomotors and PIC microcontroller. Notice the 6-V battery pack is powering the microcontroller as well as the servomotors. The battery pack is a 16-V unit using four AA batteries.

The microcontroller circuit may also be built on a small printed-circuit board that is available from Images SI Inc. (see Fig. 10.22). The robot will function for a short time using a fresh 9-V battery, it will deplete quickly. A secondary battery pack may be laid on top of the aluminum body and connected to the PC board using a power plug.

Figure 10.19 Front view of switch assembly attached to robot base.

Figure 10.20 Bottom view close-up of switch assembly.

Figure 10.23 shows the completed walker ready to run.

Microcontroller program

The 16F84 microcontroller controls the three servomotors, using just three I/O lines. This leaves 10 available I/O lines and plenty of programming space left over to improve and add to this basic walker. The program follows:

```
'Hexapod walker

'Notes
'Servomotor configuration
'Left leg(s) servomotor connected to rb4
'Right leg(s) servomotor connected to rb5
'Center tilt servomotor connected to rb6

'Pulse width out signals for following servomotors:
'Left leg (150 center) (180 forward) (120 back)
'Right leg (150 center) (120 forward) (180 back)
'Tilt (left 170) (right 130) (center 150)
```

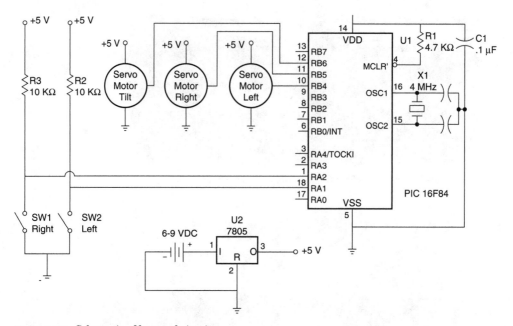

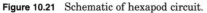

Figure 10.21 Schematic of hexapod circuit.

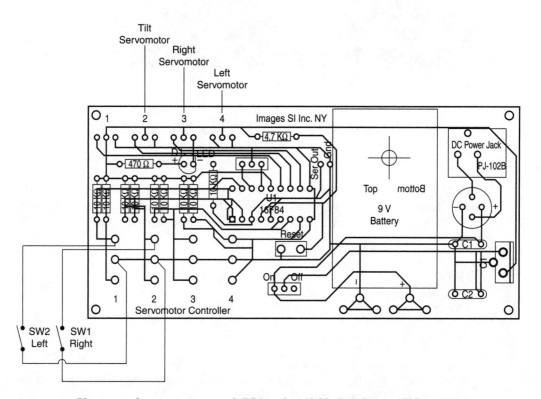

Figure 10.22 Placement of components on stock PC board available from Images SI Inc.

Figure 10.23 Finished robot.

```
'Declare variables
ls var byte            'Left servomotor pulse width
rs var byte            'Right servomotor pulse width
cs var byte            'Center servomotor pulse width
ct var byte            'Count
b0 var byte            'Count
b1 var byte            'Count

'Define variables
ls = 150
rs = 150
cs = 150

pause 250

start:

'Read forward sensors

'Front collision?
if (porta.1 = 0 && porta.2 = 0) then
   'Both left and right sensors are hit, move backward
   for b0= 1 to 3
   gosub backstep
   next
   for b0= 1 to 4
   gosub rturn
   next

endif

'Collision on right?
```

```
    if (porta.1 = 1 && porta.2 = 0) then
       'Right sensor hit, collision on right
       for b0= 1 to 2
       gosub backstep
       next
       for b0= 1 to 4
       gosub lturn
       next
    endif

    'Collision on left?
    if (porta.1 = 0 && porta.2 = 1) then
       'Left sensor hit, collision on left
       for b0= 1 to 2
       gosub backstep
       next
       for b0 = 1 to 4
       gosub rturn
       next
    endif

    'No collision keep moving forward
    if (porta.1 = 1 && porta.2 = 1) then
       'Sensors clear
       gosub frward
    endif

    goto start

    '=============================================================
    'Function subroutines
    '=============================================================

    backstep:           'Backward step
    gosub rtilt
    gosub rlf
    gosub ltilt
    gosub llf
    gosub center
    ls = 150: rs = 150
    gosub do_it
    return

    frward:             'Forward step
    gosub rtilt
    gosub rlb
    gosub ltilt
    gosub llb
    gosub center
    ls = 150: rs = 150
```

```
gosub do_it
return


lturn:              'Left turn
gosub rtilt
gosub rlb
gosub ltilt
gosub llf
gosub center
ls = 150: rs = 150
gosub do_it
return


rturn:              'Right turn
gosub rtilt
gosub rlf
gosub ltilt
gosub llb
gosub center
ls = 150: rs = 150
gosub do_it
return


'==============================================================
'Primary subroutines
'==============================================================


do_it:          'Move robot-forward, backward, left or right
for b1 = 1 to ct
pulsout portb.6, cs
pulsout portb.5, rs
pulsout portb.4, ls
pause 18
next b1


return
center:         'Center tilt servomotor
ct = 15
cs = 150
gosub do_it
return


rlf:            'Right leg forward
ct = 20
rs = 120
gosub do_it
return


rlb:            'Right leg back
ct = 20
```

```
          rs = 180
          gosub do_it
          return

          llf:               'Left leg forward
          ct = 20
          ls = 180
          gosub do_it
          return

          llb:               'Left leg back
          ct = 20
          ls = 120
          gosub do_it
          return

          rtilt:             'Right side tilt
          ct = 15
          cs = 130
          gosub do_it
          return

          ltilt:             'Left side tilt
          ct = 15
          cs = 170
          gosub do_it
          return
```

This PicBasic program provides for forward, backward, turn left, and turn right motions. Two sensors switches on the front of the robot inform the microcontroller when it has encountered an obstacle. When an obstacle is encountered, the robot steps back and turns to the left or right, depending on which side the obstacle was encountered.

The robot is provided with a right-handedness. If a front collision is detected, the robot steps back, then turns to the right and proceeds forward.

Parts List

Servomotors

Microcontrollers (16F84)

PCB

Aluminum bars

Aluminum sheets

Threaded rods and nuts (4-40)

Plastic machine screws, nuts, and washers

Available from Images SI Inc. (see Suppliers at end of book).

Chapter

11

Speech Recognition

In the near future, speech will be the method for controlling appliances, toys, tools, computers, and robotics. There is a huge commercial market waiting for this technology to mature.

Our speech recognition circuit is a stand-alone trainable speech recognition circuit that may be interfaced to control just about anything electrical (see Fig. 11.1). The interface circuit we will build in the second part of this chapter will allow this speech recognition circuit to control a variety of electrical devices such as appliances, test instruments, VCRs, TVs, and of course robots. The circuit is trained (programmed) to recognize words you want it to recognize. The unit can be trained in any language and even nonlanguages such as grunts, birdcalls, and whistles.

To be able to control and operate an appliance (computer, VCR, TV security system, etc.) or robot by speaking to it makes it easier to work with that device, while increasing the efficiency and effectiveness. At the most basic level, speech commands allow the user to perform parallel tasks (i.e., hands and eyes are busy elsewhere) while continuing to work with the computer, appliance, instrument, or robot.

The heart of the circuit is the HM2007 speech recognition integrated circuit (see Fig. 11.2). The chip provides the options of recognizing either 40 words each with a length of 0.96 s or 20 words each with a length of 1.92 s. This speech recognition circuit has a jumper setting (jumper WD on main board) that allows the user to choose either the 0.96-s word length (40-word vocabulary) or the 1.92-s word length (20-word vocabulary).

For memory the circuit uses an 8K × 8 static RAM. There is a backup memory battery for the SRAM on the main board. This battery keeps the trained words safely stored in the SRAM when the main power is turned off. The button battery lasts approximately 2 years. Without the battery backup you would have to retrain the circuit every time the circuit was switched off.

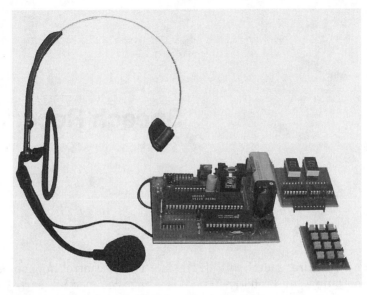

Figure 11.1 Speech recognition circuit assembled.

Figure 11.2 HM2007 integrated circuit.

The chip has two operational modes: manual mode and CPU mode. The CPU mode is implemented when it is necessary for the chip to work as a speech recognition coprocessor under a host computer. This is an attractive approach to speech recognition for computers because the job of listening to sound and recognition of command words doesn't occupy any of the main computer's CPU time. In one type of programming scenario, when the HM2007 recognizes a command, it can signal an interrupt to the host CPU and then relay the command it recognized. The HM2007 chip can be cascaded to provide a larger word recognition library.

The SR-06 circuit we are building operates in the stand-alone manual mode. As a stand-alone circuit, the speech recognition circuit doesn't require a host computer and may be integrated into other devices to add speech control.

Applications

Applications of command and control of appliances and equipment include these:

Telephone assistance systems

Data entry

Speech-controlled toys

Speech and voice recognition security systems

Robotics

Software Approach

Currently most speech recognition systems available today are software programs that run on personal computers. The software requires a compatible sound card be installed in the computer. Once activated, this software runs continuously in the background of the computer's operating system (Windows, OS/2, etc.) and any other application program.

While this speech software is impressive, it is not economically viable for manufacturers to add personal computer systems to control a washing machine or VCR. The speech recognition software steals processing power from the operating system and adds to the computer's processing tasks. Typically there is a noticeable slowdown in the operation and function of the computer when voice recognition is enabled.

Learning to Listen

We take our ability to listen for granted. For instance, we are capable of listening to one person speak among several at a party. We subconsciously filter out the extraneous conversations and sound. This filtering ability is beyond the capabilities of today's speech recognition systems.

Speech recognition is not speech understanding. Understanding the meaning of words is a higher intellectual function. The fact that a computer can respond to a vocal command does not mean it understands the command spoken. Voice recognition systems will one day have the ability to distinguish linguistic nuances and the meaning of words, to "Do what I mean, not what I say!"

Speaker-Dependent and Speaker-Independent Recognition

Speech recognition is classified into two categories, speaker-dependent and speaker-independent.

Speaker-dependent systems are trained by the individual who will be using the system. These systems are capable of achieving a high command count and better than 95 percent accuracy for word recognition. The drawback to this approach is that the system only responds accurately to the individual who trained the system. This is the most common approach employed in software for personal computers.

Speaker-independent systems are trained to respond to a word regardless of who speaks. Therefore the system must respond to a large variety of speech patterns, inflections, and enunciations of the target word. The command word count is usually lower than that of the speaker-dependent system; however, high accuracy can still be maintained within processing limits. Industrial requirements more often require speaker-independent voice systems, such as the AT&T system used in the telephone systems.

Recognition Style

Speech recognition systems have another constraint concerning the style of speech they can recognize. They are three styles of speech: isolated, connected, and continuous.

Isolated speech recognition systems can just handle words that are spoken separately. This is the most common speech recognition system available today. The user must pause between each word or command spoken. The speech recognition circuit is set up to identify isolated words of 0.96-s length.

Connected speech recognition system is a halfway point between isolated word and continuous speech recognition. It allows users to speak multiple words. The HM2007 can be set up to identify words or phrases 1.92 s in length. This reduces the word recognition vocabulary number to 20.

Continuous speech is the natural conversational speech we are used to in everyday life. It is extremely difficult for a recognizer to sift through the text as the words tend to merge together. For instance, "Hi, how are you doing?" sounds like "Hi, howyadoin." Continuous speech recognition systems are on the market and are under continual development.

Speech Recognition Circuit

The speech recognition circuit is available as a kit from Images SI Inc. You can purchase the main components, HM2007, SRAM, and printed-circuit boards separately if you like and build from scratch. The kit takes a modular approach and uses three separate printed-circuit (PC) boards. The three PC boards are the main circuit board containing the speech recognition circuit, digital display board, and keypad (see Fig. 11.3). The keypad and digital display are removable from the main circuit board. They are needed to communicate with and program the main speech recognition circuit. After the programming is accomplished, the digital display and keyboard can be removed, and the main circuit embedded into another circuit to add speech control.

Circuit construction

The schematic is shown in Fig. 11.4. You can hardwire this circuit to a breadboard if you like. I would recommend purchasing the three PCB boards that

Keypad   Display Board

Main Circuit Board

Figure 11.3 Three modular circuit boards.

are available for this project; see Parts List. When you use the PC board, the components are mounted on the top silkscreen side of the board. Begin construction by soldering the IC sockets onto the PC boards. Next mount and solder all the resistors. Now mount and solder the 3.57-MHz crystal and red LED. The long lead of the LED is positive. Next solder the capacitors and 7805 voltage regulator. Solder the seven position headers on the keypad to the main circuit board. Next solder the 10 position headers on the display board and main circuit board.

Keypad

The keypad is made up of 12 normally open (N.O.) pushbutton switches (see Fig. 11.5).

1	2	3
4	5	6
7	8	9
*	0	#
Clear		Train

To train

To train the circuit, first attach the keypad and digital display to the main circuit board (see Fig. 11.6). Next select your word length. Place a jumper on the two pin WD header on the main circuit board to select a 20-word vocabulary, each with a 2-s word length. Leave the jumper off to select a 40-word vocabulary, each with a 1-s word length. Plug in the headset microphone. When power is applied, the HM2007 checks the static RAM, outputs "00" on the digital display, and lights the red LED (READY). The circuit is in the ready

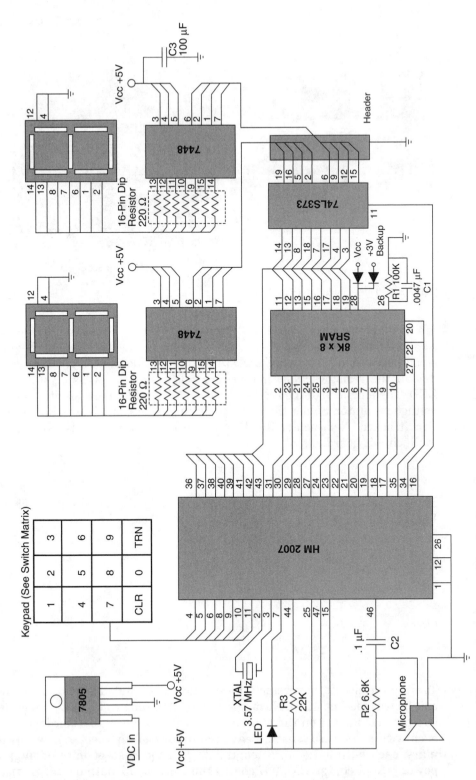

Figure 11.4 Schematic of speech recognition circuit.

KEYPAD

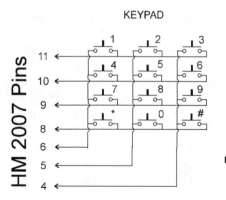

Figure 11.5 Keypad wiring.

Figure 11.6 Modular components put together for training.

mode. In the ready mode the circuit is listening for a verbal command or waiting to be trained.

To train the circuit, begin by pressing the word number you want to train on the keypad. In this exercise I am assuming you choose the 20-word vocabulary. In this mode the circuit can be trained to recognize up to 20 words. Use any numbers between 1 and 20. For example, press the number 1 to train word number 1. When you press the number(s) on the keypad, the red LED will turn off. The number pressed on the keypad is shown on the digital display. Next press the # key for train. When the # key is pressed, it signals the chip to listen for a training word, and the red LED turns back on. Now speak the word you want the circuit to recognize into the headphone microphone clearly. The LED should blink off momentarily; this is a signal that the word has been accepted.

Continue training new words in the circuit, using the procedure outlined above. Press the 2 key, then the # key to train the second word, and so on. The circuit will accept up to either 20 or 40 words, depending on the lengths of the words. You do not have to enter 20 words into memory to use the circuit. If you want, you can use as few word spaces as you require.

The procedure for training 40 words is identical, except that you can choose word numbers between 1 and 40.

Testing Recognition

The circuit is continually listening. Repeat a trained word into the microphone. The number of the word should be displayed on the digital display. For instance, if the word *directory* was trained as word number 5, then saying the word *directory* into the microphone will cause the number 5 to be displayed.

Error codes

The chip provides the following error codes.

55 = word too long

66 = word too short

77 = word no match

Clearing the trained word memory

To erase all the words in the SRAM memory (training), press 99 on the keypad and then press the * key. The display will scroll through the numbers 1 through 20 (or 1 through 40 if in 1-s word length mode) quickly, clearing out the memory.

To erase a single word space, press the number of the word you want to clear and then press the * key.

Independent Recognition System

In addition to speech commands, this circuit allows you to experiment with other facets of speech recognition technology. For instance, you can experiment with speaker-independent systems. This system is inherently speaker-dependent, meaning that the voice that trained the circuit also uses it. To experiment with speaker-independent recognition (multiuser), try the following technique. Set the WD jumper on the main circuit board to the 40-word vocabulary with a 0.96-s word length. Now we will use four word spaces for each command word. We will arrange the words so that the command words will be recognized by just decoding the least significant digit (number) on the digital display.

This is accomplished by allocating the word spaces 01, 11, 21, and 31 to the first target or command word. When the circuit is in recognition mode, we only decode the least significant digit number, in this case X1 (where X is any number from 0 to 3) to recognize the target word.

We do this for the remaining word spaces. For instance, the second target word will use word spaces 02, 12, 22, and 32. We continue in this manner until all the words are programmed.

If possible, use a different person to speak the word. This will enable the system to recognize different voices, inflections, and enunciations of the target word. The more system resources that are allocated for independent recognition, the more robust the circuit will become.

There are certain caveats to be aware of. First you are trading off word vocabulary number for speaker independence. The effective vocabulary drops from 40 words to 10 words.

The speech interface control circuit shown later may be used in this speaker-independent experimental capacity.

Voice Security System

This HM2007 wasn't designed for use in a voice security system. But this doesn't prevent you from experimenting with it for that purpose. You may want to use three or four keywords that must be spoken and recognized in sequence in order to activate a circuit that opens a lock or allows entry.

Speech Interface Control Circuit

Okay, you have a functioning speech recognition circuit, so now what? You need a method of allowing those voice commands to activate other electrical devices or functions. To do this, we need to build a universal speech interface circuit.

When designing this interface, I weighed options that I thought would make this interface useful to as many different users as possible. The first parameter I considered was how many outputs the interface should have. I decided upon 10 outputs. The second consideration was the type of output that the interface board should provide. Here was a tough choice. I had the option to make the output an active high signal that the user could use to activate or be detected. This output could be used on a TTL logic line or CMOS logic line, or to turn on a transistor switch or power relay in their circuitry.

The other option I thought of was to put 10 miniature SPDT relays on the interface board. This way the interface board could switch electric power on and off directly from the board.

The advantage of the active high output signal is cost. This board would cost much less than the interface board containing 10 relays. The advantage of the relay board is that the miniature power relays have enough current capacity to directly control small dc motors and other electric circuits.

I couldn't decide between the two approaches, so I have included both designs. You can choose which interface circuit suits you. The front ends of both circuits are identical and function in the same manner. The outputs are different and are explained separately.

Since we are controlling 10 outputs, we only need 11 commands—10 commands for active on/off switches and 1 command to turn everything off. In general, it is better if the main speech recognition board jumper (WD) is set to the 20 two-second word length option. The 20 two-second word mode has a better word recognition accuracy than the 40 one-second setting. However, the interface board will work with both modes. This makes it possible to experiment with the speaker-independent system described earlier.

The speech interface circuit needs to perform a couple of jobs. First it needs to determine when the speech recognition circuit has detected a spoken word. After a word has been detected, it must distinguish whether the word detected is a recognized command word or an unrecognized word. If the word is a recognized command word, it passes the binary information to the output. If the detected word is not a command word, it must block any change to the output.

How the circuit works

Before we can get into the nuts and bolts of how the interface circuit functions, we must look at the binary information output by the speech recognition circuit. The output of the speech recognition circuit consists of two 4-bit binary-coded decimal (BCD) numbers. This binary (BCD) information is shown on the speech circuit's two-digit digital display. Whenever a word is detected, the circuit uses the digital display to output the word number it has recognized, or else it outputs its unrecognized/error code. If the word detected is not recognized, the circuit will display one of the following error codes:

55 = word too long

66 = word too short

77 = word no match

Our interface design incorporates a PIC microcontroller (see Fig. 11.7 or 11.8). A preprogrammed microcontroller's (16F84) first job is to determine if a word has been spoken. To do this, we use an LM339 comparator. A reference voltage for the comparator is generated using a voltage divider made up of

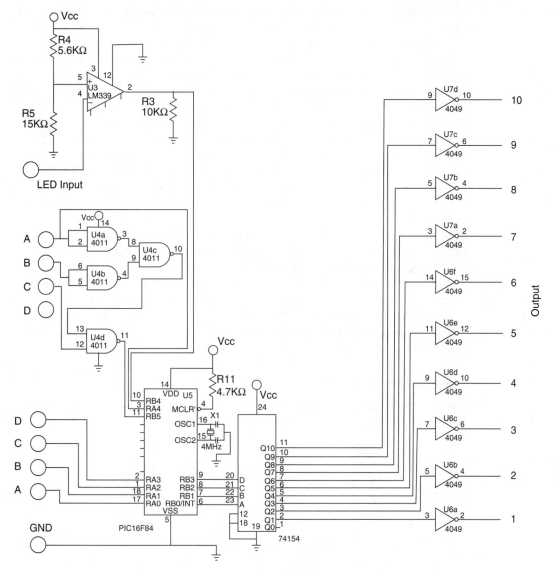

Figure 11.7 Speech recognition interface (active high outputs) SRI-03.

resistors R4 and R5. The reference voltage is placed on pin 5 of the comparator. Pin 4 of the comparator is connected to the LED lead on the speech recognition circuit. Whenever a word is detected, the LED blinks off momentarily. The output of the comparator (pin 2) is connected to pin 10 (RB4) of the 16F84 microcontroller. The output of the comparator (pin 2) is usually high (+5 V). When a word is detected, the output (pin 2) drops to ground momentarily. The microcontroller monitors this line to determine when a word has been detected.

Once a word has been detected, it is necessary for the interface to read the BCD output from the speech recognition circuit. By using the high- and low-digit BCD nibbles, it's possible to distinguish trained target words. To do so, the interface must distinguish the error codes 55, 66, and 77 from trained words numbered 5, 6, and 7. To accomplish this, the interface circuit uses four NAND gates off the 4011 integrated circuit. The NAND gates are connected to the high-digit nibble. If the high-digit BCD nibble has the equivalent word numbers of 5, 6, or 7, the output from the four NAND gates is low. The output from the four NAND gates is connected to pin 11 (RB5) of the 16F84. The 16F84 reads this pin to determine if the high-digit nibble is a 5, 6, or 7 (0 V or ground). If these numbers are not displayed, the output of the NAND gates is high (+5 V).

So far our circuit can tell when a word has been detected and if the resulting word is an error code. If the output of the speech recognition circuit is an error code, nothing else happens; the microcontroller loops back to the beginning of the program, waiting for another word detection. On the other hand, if a word is detected and it is not an error code, the microcontroller passes the low-digit number through to the 74HC154 (4- to 16-line decoder) IC. The 74HCT154 decoder reads the binary number passed to it and brings the corresponding pin equivalent to that number low.

PIC 16F84 microcontroller program

The PIC 16F84 used in both interface circuits contains the following PicBasic program:

```
'Speech recognition interface program
symbol porta = 5
symbol trisa = 133
symbol portb = 6
symbol trisb = 134
poke trisa, 255
poke trisb, 240
start:
peek portb, b0
if bit4 = 0 then trigger   'Trigger enabled, read speech recognition
                             circuit
goto start                 'Repeat
trigger:
pause 500                  'Wait .5 second
peek portb, b0             'Read bcd number
if bit5 = 1 then send      'Output number
goto start                 'Repeat
send:
peek porta, b0             'Read port a
if bit4 = 1 then eleven    'Is the number 11
poke portb, b0             'Output number
```

```
goto start              'Repeat
eleven:
if bit0 = 0 then ten
poke portb, 11
goto start              'Repeat
ten:
poke portb, 10
goto start              'Repeat
end
```

Active high output

The outputs from the 74HCT154 each pass through a 4049 inverting buffer to supply a 15-Vdc active high output signal.

SPDT relay output

In Fig. 11.8, the front end of the circuit is identical to Fig. 11.7. The changes are seen in the back end of the circuit. The active low output signals from the 74HCT154 each connect to one of the 10 PNP transistors, each of which controls a corresponding relay. Each relay has a normally open (N.O.) switch and normally closed (N.C.) switch. The relay switches are rated at 124 V ac at 0.5 A or 24 V dc at 1 A. The relay itself consumes approximately 30 mA of current when it is turned on.

Circuit Construction

There is nothing critical about the circuit construction. The circuit may be wired point to point on a breadboard, if you like. Printed-circuit boards make the construction easier and are available as kits from Images SI Inc.

The only component that needs special notice is the 10-pin female header. If you are not using the PC boards from the kit, you must follow the schematic and wire the 10-pin female header exactly; or else the interface will not be receiving the signals it expects, and the unit will fail.

Programming the Speech Recognition Circuit: Training, Testing, and Retraining

Program the speech recognition circuit per the directions given previously. Choose the words you want to use to control the 10 electrical relays or outputs. To turn off all electrical outputs on the interface, train word number 11 as *stop, end,* or *quit.*

Before you connect the interface to any circuit, repeat all the trained words into the microphone. The corresponding word number will be displayed on the digital display. You should achieve recognition accuracy of better than 95 percent. If the circuit continually confuses two training words, try retraining one of the words. To retrain a word, press the word number, using the keypad; the word number will be displayed on the digital display.

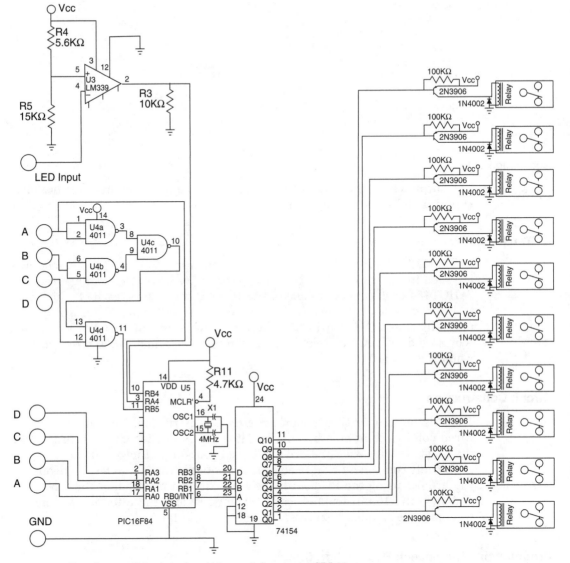

Figure 11.8 Speech recognition interface (relay switch outputs) SRI-02.

Press the T (training) key, and say the word into the microphone. If the circuit still confuses the two words, you may have to change one of the suggested words.

Once you are satisfied with the accuracy, remove the digital display board and the keypad. Next connect the speech interface board to the 10-pin header used for the digital display, and you're ready to go.

Figure 11.9   Finished speech recognition board SRI-02.

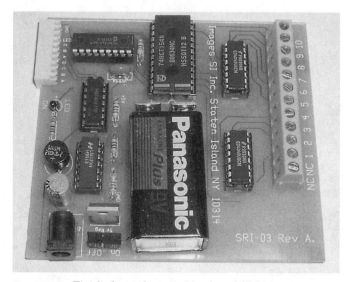

Figure 11.10   Finished speech recognition board SRI-03.

SRI-02 and SRI-03 Interfaces

The SRI-02 and SRI-03 built from kits available from Images SI Inc. are shown in Figs. 11.9 and 11.10, respectively. Once the speech recognition circuit is programmed, the speech recognition interfaces may be plugged into the display board output on the main speech recognition board and used. Figure 11.11 shows the SRI-02 connected to the speech recognition board, and Fig. 11.12 shows the SRI-03 connected to the speech recognition board.

Robot Control

The speech recognition circuit uses a headphone microphone. For mobile operation one needs to add a wireless microphone. There are a number of methods of implementing wireless control.

The simplest method is to add a suitable microphone to the main circuit board and acoustically couple it to the output of a radio receiver or walkie-talkie. You would use the matching walkie-talkie to give voice commands. When using this method, you should train the circuit by using your walkie-talkies and acoustic coupling.

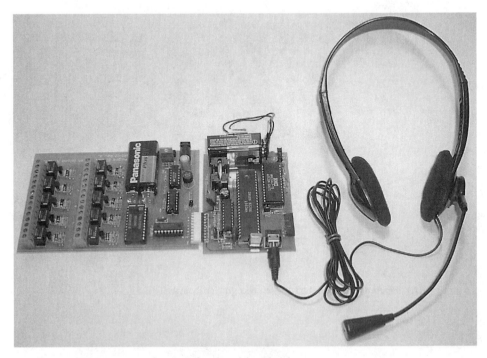

Figure 11.11 SRI-02 connected to speech recognition circuit.

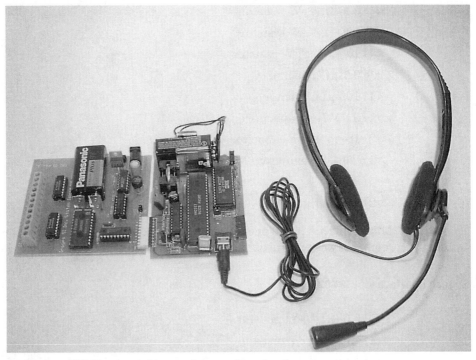

Figure 11.12 SRI-03 connected to speech recognition circuit.

Parts List

Speech recognition kit (SR-06)

(1) Speech recognition IC (HM2007)

(1) 8K static RAM (6264)

(1) Octal latch (74LS373)

(1) Display chip (74LS48)

(1) 3.57-MHz crystal

(12) PC-mounted N.O. switches

(2) Seven-segmented displays (MAN74)

(1) Headset microphone

(1) 9-V battery clip

(1) Coin battery holder (2032)

(1) PC-mounted microphone jack

(1) 22-kΩ, $^1/_4$-W resistor

(1) 6.8-kΩ, $^1/_4$-W resistor

(1) 330-Ω, $\frac{1}{4}$-W resistor

(8) 220-Ω, $\frac{1}{4}$-W resistor

(1) 100-kΩ, $\frac{1}{4}$-W resistor

(1) 0.1-μf capacitor

(1) 100-μf capacitor

(1) 0.0047-μf capacitor

(2) 10- to 22-pF capacitor

(1) Voltage regulator (7805)

(1) LED

(2) 1N914 diode

Miscellaneous items needed include PC boards, IC sockets, headers (male and female), two- and three-pin connectors, jumpers.

Speech interface kit (SRI-02)

(1) 5.6-kΩ, $\frac{1}{4}$-W resistor

(1) 15-kΩ, $\frac{1}{4}$-W resistor

(1) 10-kΩ, $\frac{1}{4}$-W resistor

(10) 100-kΩ, $\frac{1}{4}$-W resistor

(10) Diodes (1N4002)

(1) Comparator (LM339)

(1) 4011 CMOS NAND

(1) 74154 IC

(1) PIC 16F84 microcontroller*

(10) Omron G5V-1 relays

Miscellaneous items needed include PC board, 10-pin female header, 9-V battery clips, and a 7805 regulator.

Speech interface kit (SRI-03)

(1) 5.6-kΩ, $\frac{1}{4}$-W resistor

(1) 15-kΩ, $\frac{1}{4}$-W resistor

(1) 10-kΩ, $\frac{1}{4}$-W resistor

(10) 100-kΩ, $\frac{1}{4}$-W resistor

---

*Preprogrammed 16F84 available separately for $10.00 from Images SI Inc.

(10) Diodes (1N4002)

(1) Comparator (LM339)

(1) 4011 CMOS NAND

(1) 74154 IC

(1) PIC 16F84 microcontroller*

(2) Inverting buffers (4049)

Miscellaneous items needed include PC board, 10-pin female header, 9-V battery clips, and a 7805 regulator.

Speech recognition and interface kits (all components including preprogrammed 16F84 and PCB) available from Images SI Inc. (see Suppliers at end of book):

Speech recognition kit (SR-06)	$79.95
Speech interface kit (SRI-03)	$89.95
Speech interface kit (relay) (SRI-02)	$159.95

*Preprogrammed 16F84 available separately for $10.00 from Images SI Inc.

12

Robotic Arm

Servomotor Building Blocks for Robotics

The servomotor brackets discussed in this chapter will allow you to create various servomotor robots and projects.

Servomotors are ideal for powering robots. They are readily available in many sizes, are inexpensive, provide powerful torque for their size and weight, and are positional. The output shafts on most hobby servomotors are guaranteed positional between 0° and 90°. Most servomotors' output shaft range extends past 90°, coming close to 180°.

The servomotor bracket components are shown in Fig. 12.1. Each of the aluminum U brackets that make up the assembly has multiple holes for connecting a standard HiTec servomotor horn as well as bottom and top holes for connecting U brackets and assemblies to one another.

The servomotor horns used on these servomotor brackets are included with all the compatible HiTec servomotors, such as HS-322, HS-425, HS-475, and HS-35645. These brackets may also be used with similar-size Futaba servomotors, but you may have to purchase the horns separately.

Each servomotor bracket assembly consists of the following components: two aluminum U brackets, labeled A and B, one binding head post screw, four 6-32 plastic machine screws with nuts, and four sheet metal screws for mounting a servomotor horn. When assembled with a compatible servomotor (see Fig. 12.2), the bracket becomes a modular component that may be attached to other brackets and components. The bracket allows the top and bottom components to swivel along the axis of the servomotor's shaft (see Fig. 12.3).

By connecting multiple servomotors using the brackets, you can create a variety of robotic designs. In this chapter we will use the brackets to create a

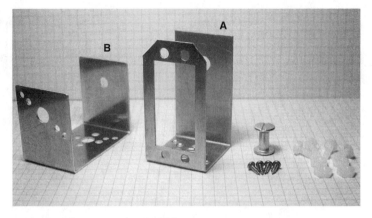

Figure 12.1   Servomotor bracket kit.

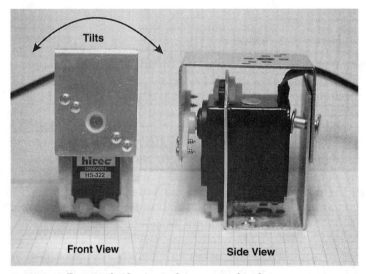

Figure 12.2   Front and side views of servomotor bracket.

five-servomotor robotic arm. In Chap. 13 we use these same brackets to create a bipedal walker robot.

The bottom and top have multiple holes for attaching other brackets or servomotor horns (see Fig. 12.4).

Basic Servomotor Bracket Assembly

To assemble a servomotor bracket, begin by placing the binding post through the back hole on part a (see Fig. 12.5). Next place servomotor into the A bracket, as shown in Fig. 12.6. Attach the servomotor using 6-32 × $^3/_8$-in-long machine screws and nuts (see Fig. 12.7). Notice the servomotor's horn has

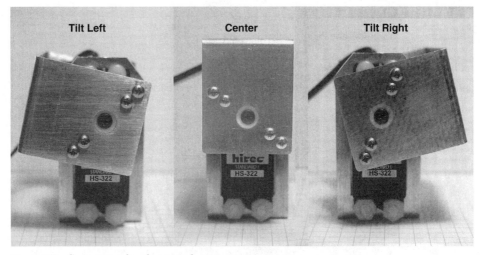

Figure 12.3 Servomotor bracket travel.

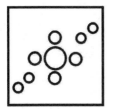
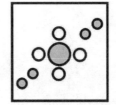
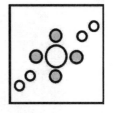

Bracket Holes   Horn-Mounting Holes   Bracket-to-Bracket Holes

Figure 12.4 Diagram of top and bottom mounting holes in the A and B brackets.

Figure 12.5 A bracket with binding screw.

Figure 12.6 Side view of placing servomotor in A bracket.

Figure 12.7 A bracket with servomotor attached with plastic screws and nuts.

been removed from the servomotor. To secure the screws at the bottom two positions of the servomotor, place the screw through the hole from the inside of the bracket. It helps if you have a small screwdriver to hold the screw in place. Then the plastic nuts are chased down on the screws from the outside of the bracket (see Fig. 12.7).

The servomotor horn (see Fig. 12.8), is attached to the side holes on the B bracket (see Fig. 12.9).

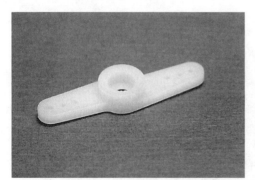

Figure 12.8 HiTec servomotor horn.

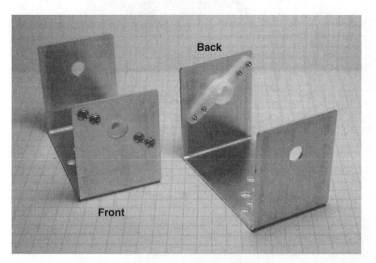

Figure 12.9 B bracket with servomotor horn attached.

To place the servomotor secure in bracket A into its mating part bracket B, slip the end of the binding-held post through the hole in the mating part (see Fig. 12.10). Next slip the servomotor's spindle into the horn (see Fig. 12.11). Finished assembly is shown in Fig. 12.12.

Assembling Multiple-Servomotor Assemblies

When you are using multiple-servomotor assemblies, it is essential to preplan how the servomotors will be connected. When two or more servomotors assemblies are connected, the connecting brackets of the joints should be pre-assembled (see Fig. 12.13). The brackets may be orientated to one another in a number of ways, depending upon your design.

The top and bottom brackets of each assembly are connected to one another by four 6-32 × $^3/_8$-in-long plastic machine screws and eight plastic hex nuts. The screws are inserted though the top bracket holes. Hex nuts

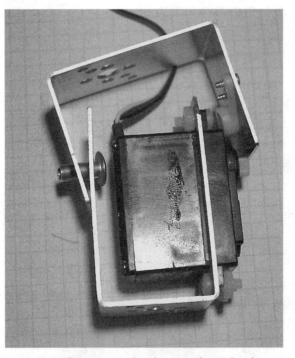

Figure 12.10 Bringing top bracket onto lower bracket to assemble.

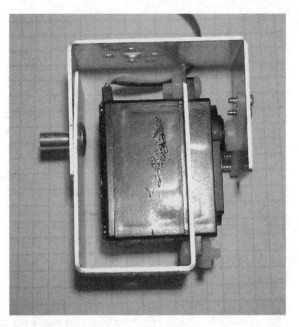

Figure 12.11 Side view showing horn assembly connected to servomotor.

Figure 12.12 Stand-alone servomotor bracket assembly.

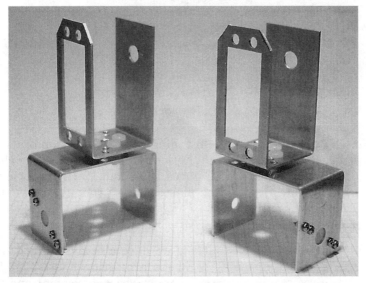

Figure 12.13 Two different bracket assemblies.

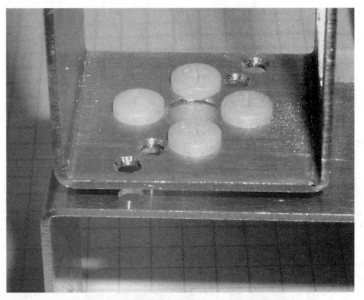

Figure 12.14   Close-up top view of two assembled brackets.

Figure 12.15   Close-up side view of two assembled brackets.

are chased down, securing the machine screws to the top bracket. The second bracket is then attached to the screws, and hex nuts are chased down, securing the bottom bracket. Figures 12.14 and 12.15 are close-up pictures of the top and side views of the plastic screws connecting two brackets.

Building a Five-Servomotor Robotic Arm

Aside from the servomotor brackets we have already outlined, we need one other specialized component—a robotic arm gripper (see Fig. 12.16). This gripper requires two servomotors, one for wrist movement and the other to open and close the gripper fingers. The gripper fingers can accommodate objects up to about 1.0 in (25 mm).

Figure 12.16 Robotic arm gripper.

The robotic arm uses five servomotors: four HiTec HS-322 HD servomotors and one HS-475 HB servomotor. The HS-475 servomotor has 50 percent more torque than the HS-322 and is used in the second position up from the bottom (or base) servomotor on the robotic arm. This particular servomotor requires the greatest torque in order to lift the arm and any object the arm is holding.

Figure 12.17 shows how the servomotors are attached to the gripper. Assemble one part A and B bracket, as shown in Fig. 12.18. Attach a servomotor to the A portion of the bracket; this will be the wrist servomotor. The wrist servomotor motor is attached to the gripper first. Remove the servomotor horn from the servomotor, if you haven't done so already, and put the horn screw to the side; we will need it. Center the wrist servomotor, using the centering servomotor circuit described later in this chapter or at the end of Chap. 6. With power applied to the servomotor from the centering circuit, place the servomotor into the wrist position. Replace the horn screw removed earlier, and tighten the servomotor horn screw. Remove power from the servomotor.

Next position the gripper fingers in midposition. Center the finger servomotor, using the centering circuit as before. Position the finger servomotor in the finger position. Tighten the horn servomotor screw, then back off the screw to unbind the fingers. When you are finished, the subassembly should look like Fig. 12.19.

To finish up the arm, assemble an A and B component, as shown in Fig. 12.20. Next we require two more A bracket components. One A bracket component has a servomotor horn attached to its bottom holes, and the other A bracket component has a servomotor attached and is laid on its back as a base (see Fig. 12.21). The two brackets are assembled as shown in Fig. 12.22. When you assemble the base, center the bottom servomotor before attaching the upper A bracket. This forms the base of the robotic arm. To secure the base to a platform, four holes are drilled in the bottom bracket (see Fig. 12.23). Only two drill locations are shown on the bottom. Drill two similar holes at the top. To prevent the A bracket from bending with the weight of the robotic arm when it is assembled, place a spacer made of wood, plastic, or metal as shown in Fig. 12.23. The base assembly is secured to a square piece

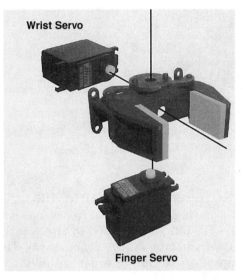

Wrist Servo

Finger Servo

Figure 12.17 Diagram showing how servomotor assembles to gripper.

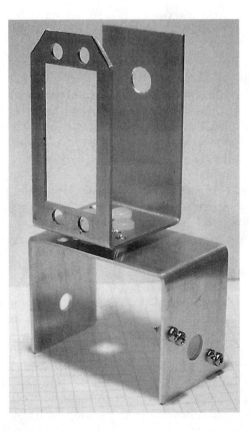

Figure 12.18 Assembled brackets for gripper.

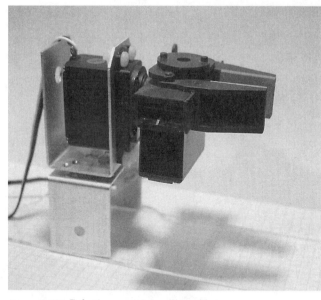

Figure 12.19 Robotic arm gripper assembly.

Figure 12.20 Assembled middle bracket for robotic arm.

Figure 12.21   Bottom brackets for robotic arm.

Figure 12.22   Assembled bottom brackets for robotic arm.

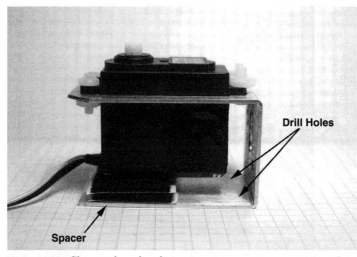

Figure 12.23 Close-up base bracket.

of wood or metal to provide a good base that doesn't topple when the robotic arm moves and lifts objects.

The two middle servomotors are assembled onto the base, and the servogripper is attached to the top, completing the robotic arm (see Figs. 12.24 and 12.25).

Servomotors

Servomotors are relatively easy to control using PIC microcontrollers. If you remember, servomotors were introduced in Chap. 6. In Chap. 6 we just described the basic function of a servomotor; now we will review in a little greater detail.

Servomotors are geared dc motors with a positional feedback control that allows the shaft (rotor) to be rotated and positioned accurately. When a control signal is being fed to the servomotor, the servomotor's shaft rotates to the position specified by the control signal. The positioning control is a dynamic feedback loop, meaning that if you forcibly rotate the servomotor's shaft away from its control signal command position, the servomotor circuitry will read this as a position error and will increase its torque in an attempt to rotate the shaft back to its command position.

Hobby servomotor specifications usually state that the shaft can be positioned through a minimum range of 90° (±45°). In reality this range can be extended closer to 180° (±90°) by adjusting the position control signal described in a moment.

There are three wire leads to a hobby servomotor. Two leads are for power 15 V (red wire) and ground (black wire). The third lead (yellow or white wire) feeds a position control signal to the motor.

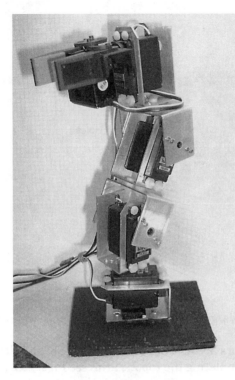

Figure 12.24 Five-servomotor robotic arm (left view).

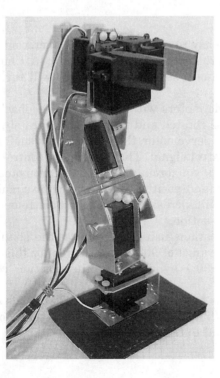

Figure 12.25 Five-servomotor robotic arm (right view).

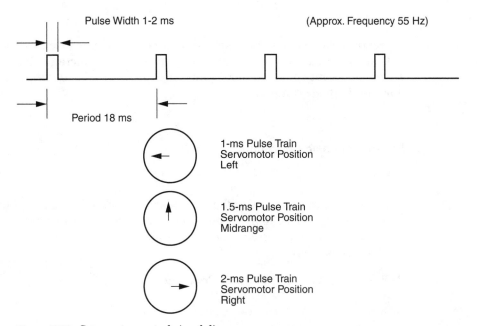

Figure 12.26 Servomotor control signal diagram.

The position control signal is a single variable-width pulse. The pulse width typically varies between 1 and 2 ms. The width of the pulse controls the position of the servomotor shaft. Figure 12.26 illustrates the relationship of pulse width to servomotor position. A 1-ms pulse rotates the shaft to the extreme counterclockwise (CCW) position (−45°). A 1.5-ms pulse places the shaft in a neutral midpoint position (0°). A 2-ms pulse rotates the shaft to the extreme CW position (+45°).

The pulse width signal is sent to the servomotor approximately 55 times per second (55 Hz).

By extending our pulse width past the typical parameters, a 1- to 2-ms pulse width, we can extend the rotational position of the servomotor's shaft. In many cases close to 180° positioning control is possible. However, care must be exercised not to provide a control signal to the servomotor that will attempt to rotate the shaft too far, where the shaft will push against its internal stop. As mentioned previously, the position feedback control is dynamic, and the servomotor will increase its torque (and increase its current consumption) to rotate the shaft into position, placing as much force as possible against its internal stop. This will create unnecessary strain on the internal gears and motor, decreasing its working life considerably.

Servomotor controllers

Our servomotor controllers use the PicBasic and PicBasic Pro `pulsout` command. The command format is as follows:

```
pulsout pin, period
```

The pulsout command generates a pulse on the pin specified for the period of time specified. The time is in 10-μs (microsecond) increments. So to send a 1.5-ms pulse out on port B pin 0, you could use one of the following command(s). For the PicBasic compiler:

```
pulsout 0, 150
```

For the PicBasic Pro compiler:

```
pulsout portb.0, 150
```

This pulsout command will put the servomotor shaft into its center position. The only things missing are a delay and loop-back lines to send the pulsout signal to the servomotor 55 times per second. So a complete center servomotor program is as follows:

PicBasic program	PicBasic Pro program
start:	start:
pulsout 0, 150	pulsout portb.0, 150
pause 18	pause 18
goto start	goto start

The schematic for a basic servomotor circuit is shown in Fig. 12.27. If you prototype servomotor circuits on a solderless breadboard, a servomotor connector (see Fig. 12.28) makes connecting a servomotor to the breadboard easy.

Although this centering servomotor circuit may appear to be useless, it is not. In most cases when building a servomotor device or robot, you want to center the servomotor to a known (center) position before attaching any hardware. This centering technique is used before attaching the wheel assembly to the steering servomotor when you are constructing Walter's turtle (see Chaps. 8 and 10 among others).

Simple servomotor controller

This second servomotor circuit (see Fig. 12.29), allows us to control the servomotor by using a single-pole double-throw (SPDT) switch. This particular SPDT switch has a center-off position that is critical to proper operation of this circuit. Pushing the switch up will rotate the servomotor in a clockwise rotation. In the center position the servomotor stops and holds its position. Pushing the switch in the down position will rotate the servomotor in the counterclockwise direction.

The following two programs for the simple servomotor controller are the basis for the programming for the four- and five-servomotor controllers. In general, when you are programming the PIC microcontrollers, make sure the watchdog timer is disabled.

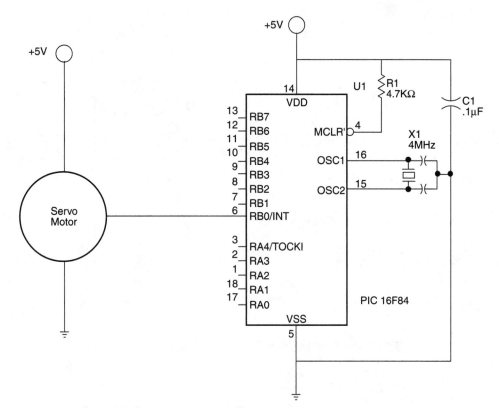

Figure 12.27   Centering the servomotor controller circuit.

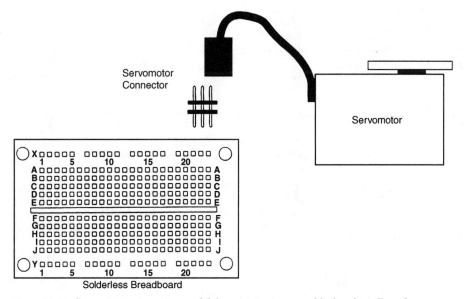

Figure 12.28   Servomotor connector useful for prototyping on solderless breadboards.

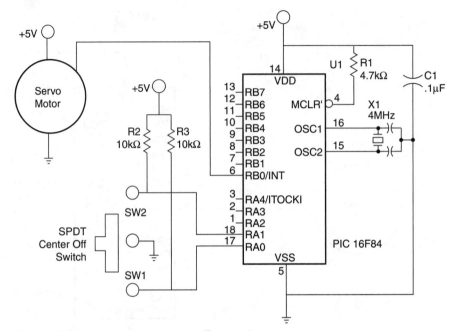

Figure 12.29 Primary servomotor controller circuit.

```
'PicBasic Pro program
'Manual control of servomotor using SPDT switch
'Use b1 to hold pulse width variable for servo 1
'Declare variables

b1 var byte

'Initialize variables

b1 = 150                                'Start servo 1 at center position

start:

'Output servomotor position
    pulsout portb.0, b1                 'Send current servo 1 position out

'Check for switch closures

    if porta.0 = 0 then left1           'Is sw1 left active?
    if porta.1 = 0 then right1          'Is sw1 right active?

'Routine to adjust pause value (nom 18) to generate approx 50 Hz update

pause 18
    goto start
```

```
'Routines for servomotor 1
left1:
    b1 = b1 + 1                        'Increase the pulse width
    if b1 > 254 then max1              'Maximum 2.54 milliseconds
    goto start
right1:
    b1 = b1 - 1                        'Decrease the pulse width
    if b1 < 75 then min1               'Minimum .75 millisecond
    goto start
max1:
    b1 = 254                           'Cap max b1 at 2.54 milliseconds
    goto start
min1:
    b1 = 75                            'Cap min b1 at .75 millisecond
    goto start


'PicBasic program
'Manual control of servomotor using SPDT switch
'Use b1 to hold pulse width variable for servo 1
'Declare variables

'Initialize variables

symbol porta = 6

b1 = 150                              'Start servo 1 at center position

start:

'Output servomotor position

    pulsout 0, b1                     'Send current servo 1 position out

'Check for switch closures

peek porta, b0
    if bit0 = 0 then left1            'Is sw1 left active?
    if bit1 = 0 then right1           'Is sw1 right active?

'Routine to adjust pause value (nom 18) to generate approx 55 Hz update
pause 18
    goto start

'Routines for servomotor 1
left1:
    b1 = b1 + 1                        'Increase the pulse width
    if b1 > 254 then max1              'Maximum 2.54 milliseconds
  goto start
right1:
```

```
        b1 = b1 - 1                  'Decrease the pulse width
        if b1 < 75 then min1         'Minimum .75 millisecond
        goto start
    max1:
        b1 = 254                     'Cap max b1 at 2.54 milliseconds
        goto start
    min1:
        b1 = 75                      'Cap min b1 at .75 millisecond
        goto start
```

Four- and Five-Servomotor Controllers

The previous schematic is the basic building block used in the four- and five-servomotor controller. Figure 12.30 shows the four-servomotor controller. This may be purchased as a kit from Images SI Inc., or you can hardwire the circuit and program the chip yourself.

```
'PicBasic Pro program
'Manual control of four servomotors using 4 SPDT switches
'Microcontroller PIC 16f84

'Declare variables

b0 var word      'Variable for pause routine.
b1 var byte      'Use b1 to hold pulse width variable for servo 1
b2 var byte      'Use B2 to hold pulse width variable for servo 2
```

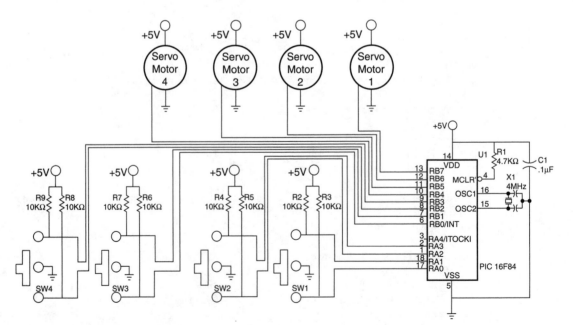

Figure 12.30 Schematic of four-servomotor controller.

```
b3 var byte      'Use b3 to hold pulse width variable for servo 3
b4 var byte      'Use b4 to hold pulse width variable for servo 4
b5 var byte      'Variable for pause routine
'Initialize servomotor variables

b1 = 150         'Start up position servo 1
b2 = 150         'Start up position servo 2
b3 = 150         'Start up position servo 3
b4 = 150         'Start up position servo 4

start:

'Output servomotor position

    pulsout portb.7, b1      'Send current servo 1 position out
    pulsout portb.6, b2      'Send current servo 2 position out
    pulsout portb.5, b3      'Send current servo 3 position out
    pulsout portb.4, b4      'Send current servo 4 position out

'Check for switch closures

    if porta.0 = 0 then left1    'Is sw1 left active?
    if porta.1 = 0 then right1   'Is sw1 right active?
    if porta.2 = 0 then left2    'Is sw2 left active?
    if porta.3 = 0 then right2   'Is sw2 right active?
    if portb.0 = 0 then left3    'Is sw3 left active?
    if portb.1 = 0 then right3   'Is sw3 right active?
    if portb.2 = 0 then left4    'Is sw4 left active?
    if portb.3 = 0 then right4   'Is sw4 right active?

'Routine to adjust pause value (nom 18) to generate approx 50 Hz update

b0 = b1 + b2 + b3 + b4
b5 = b0/100
b0 = 15 - b5

pause b0

    goto start

'Routines for servomotor 1
left1:
    b1 = b1 + 1              'Increase the pulse width
    if b1 > 254 then max1    'Maximum 2.54 milliseconds
    goto start
right1:
    b1 = b1 - 1              'Decrease the pulse width
    if b1 < 75 then min1     'Minimum .75 millisecond
    goto start
max1:
```

```
        b1 = 254                    'Cap max b1 at 2.54 milliseconds
        goto start
min1:
        b1 = 75                     'Cap min b1 at .75 millisecond
        goto start

'Routines for servomotor 2
left2:
        b2 = b2 + 1                 'Increase the pulse width
        if b2 > 254 then max2       'Maximum 2.54 milliseconds
        goto start
right2:
        b2 = b2 - 1                 'Decrease the pulse width
        if b2 < 75 then min2        'Minimum .75 millisecond
        goto start
max2:
        b2 = 254                    'Cap max b2 at 2.54 milliseconds
        goto start
min2:
        b2 = 75                     'Cap min b2 at .75 millisecond
        goto start

'Routines for servomotor 3
left3:
        b3 = b3 + 1                 'Increase the pulse width
        if b3 > 254 then max3       'Maximum 2.54 milliseconds
        goto start
right3:
        b3 = b3 - 1                 'Decrease the pulse width
        if b3 < 75 then min3        'Minimum .75 millisecond
        goto start
max3:
        b3 = 254                    'Cap max b3 at 2.54 milliseconds
        goto start
min3:
        b3 = 75                     'Cap min b3 at .75 millisecond
        goto start

'Routines for servomotor 4
left4:
        b4 = b4 + 1                 'Increase the pulse width
        if b4 > 254 then max4       'Maximum 2.54 milliseconds
        goto start
right4:
        b4 = b4 - 1                 'Decrease the pulse width
        if b4 < 75 then min4        'Minimum .75 millisecond
        goto start
max4:
        b4 = 254                    'Cap max b4 at 2.54 milliseconds
        goto start
min4:
```

```
        b4 = 75                          'Cap min b4 at .75 millisecond
        goto start
    end
```

Figure 12.31 is a photograph of the completed four-servomotor kit. The circuit board for this kit was used as the main circuit board for the turtle robot in Chap. 8. Figure 12.32 is a schematic for the five-servomotor controller. This circuit is suitable for controlling our five-servomotor robotic arm.

When you program the 16F873 with the five-servomotor controller program, make sure the watchdog timer is disabled and the brownout reset is

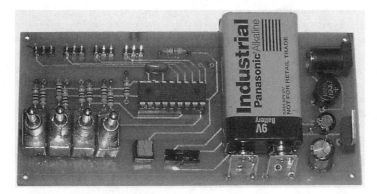

Figure 12.31 Assembled four-servomotor controller kit.

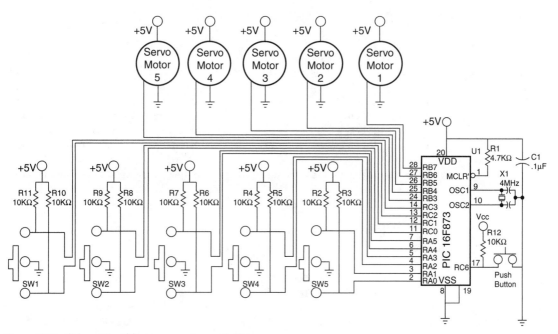

Figure 12.32 Schematic of five-servomotor controller.

also disabled. If the brownout reset is not disabled, the circuit may automatically reset whenever a servomotor draws enough current to make the supply voltage dip momentarily. This is not what you want to happen in the middle of a robotic arm operation, so make sure that configuration bit is disabled.

These configuration bits are easy to set when you use the EPIC Programmer. Simply go to the Configuration pull-down menu and disable these options.

```
'PicBasic Pro program for five-servomotor controller
'Manual control of five servomotors using 5 SPDT switches
'Microcontroller PIC 16f873
adcon1 = 7                 'Set port a to digital I/O
'Declare variables
b0 var byte                'Use b0 as hold pulse width variable for servo 1
b1 var byte                'Use b1 to hold pulse width variable for servo 2
b2 var byte                'Use b2 to hold pulse width variable for servo 3
b3 var byte                'Use b3 to hold pulse width variable for servo 4
b4 var byte                'Use b4 to hold pulse width variable for servo 5
b6 var byte                'Variable for pause routine
b7 var word                'Variable for pause routine


'Initialize servomotor variables

b0 = 150                   'Start up position servo 1
b1 = 150                   'Start up position servo 2
b2 = 150                   'Start up position servo 3
b3 = 150                   'Start up position servo 4
b4 = 150                   'Start up position servo 5


start:

'Output servomotor position
portb = 0                             'Prevents potential signal inversion on reset
    pulsout portb.7, b0               'Send current servo 1 position out
    pulsout portb.6, b1               'Send current servo 2 position out
    pulsout portb.5, b2               'Send current servo 3 position out
    pulsout portb.4, b3               'Send current servo 4 position out
    pulsout portb.3, b4               'Send current servo 5 position out

'Routine to adjust pause value (nom 18) to generate approx 50 Hz update

b7 = b0 + b1 + b2 + b3 + b4
b6 = b7/100
b7 = 15 - b6
pause b7

'Check for switch closures
    if portc.3 = 0 then left1         'Is sw1 left active?
    if portc.2 = 0 then right1        'Is sw1 right active?
    if portc.1 = 0 then left2         'Is sw2 left active?
```

```
    if portc.0 = 0 then right2          'Is sw2 right active?
    if porta.5 = 0 then left3           'Is sw3 left active?
    if porta.4 = 0 then right3          'Is sw3 right active?
    if porta.3 = 0 then left4           'Is sw4 left active?
    if porta.2 = 0 then right4          'Is sw4 right active?
    if porta.1 = 0 then left5           'Is sw5 left active?
    if porta.0 = 0 then right5          'Is sw5 right active?

    goto start

'Routines for servomotor 1
left1:
    b0 = b0 + 1                 'Increase the pulse width
    if b0 > 254 then max0       'Maximum 2.54 milliseconds
    goto start
right1:
    b0 = b0 - 1                 'Decrease the pulse width
    if b0 < 75 then min0        'Minimum .75 millisecond
    goto start
max0:
    b0 = 254                    'Cap max b1 at 2.54 milliseconds
    goto start
min0:
    b0 = 75                     'Cap min b1 at .75 millisecond
    goto start

'Routines for servomotor 2
left2:
    b1 = b1 + 1                 'Increase the pulse width
    if b1 > 254 then max1       'Maximum 2.54 milliseconds
    goto start
right2:
    b1 = b1 - 1                 'Decrease the pulse width
    if b1 < 75 then min1        'Minimum .75 millisecond
    goto start
max1:
    b1 = 254                    'Cap max b1 at 2.54 milliseconds
    goto start
min1:
    b1 = 75                     'Cap min b1 at .75 millisecond
    goto start

'Routines for servomotor 3
left3:
    b2 = b2 + 1                 'Increase the pulse width
    if b2 > 254 then max2       'Maximum 2.54 milliseconds
    goto start
right3:
    b2 = b2 - 1                 'Decrease the pulse width
    if b2 < 75 then min2        'Minimum .75 millisecond
```

```
      goto start
max2:
   b2 = 254                       'Cap max b2 at 2.54 milliseconds
   goto start
min2:
   b2 = 75                        'Cap min b2 at .75 millisecond
   goto start


'Routines for servomotor 4
left4:
   b3 = b3 + 1                    'Increase the pulse width
   if b3 > 254 then max3          'Maximum 2.54 milliseconds
   goto start
right4:
   b3 = b3 - 1                    'Decrease the pulse width
   if b3 < 75 then min3           'Minimum .75 millisecond
   goto start
max3:
   b3 = 254                       'Cap max b3 at 2.54 milliseconds
   goto start
min3:
   b3 = 75                        'Cap min b3 at .75 millisecond
   goto start


'Routines for servomotor 5
left5:
   b4 = b4 + 1                    'Increase the pulse width
   if b4 > 254 then max4          'Maximum 2.54 milliseconds
   goto start
right5:
   b4 = b4 - 1                    'Decrease the pulse width
   if b4 < 75 then min4           'Minimum .75 millisecond
   goto start
max4:
   b4 = 254                       'Cap max b4 at 2.54 milliseconds
   goto start
min4:
   b4 = 75                        'Cap min b4 at .75 millisecond
   goto start
end
```

Figure 12.33 is a photograph of the five-servomotor controller.

The robotic arm servomotors can plug right onto the three position headers on the main board. However, to separate the control board from the robotic arm, I used five 24-in servomotor extensions. Once wired, each SPDT switch controls one robotic arm servomotor (see Fig. 12.34).

When using the robotic arm, I noticed the arm move too quickly for me to perform fine movements. So to slow it down, I added a delay routine. This following program is identical to the above program, with the exception of the delay routine(s).

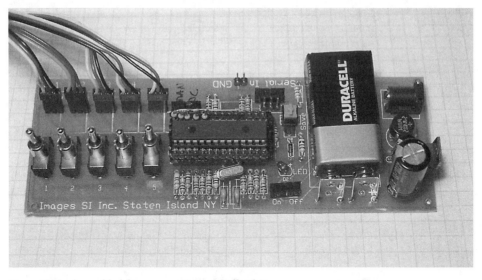

Figure 12.33 Assembled five-servomotor controller kit.

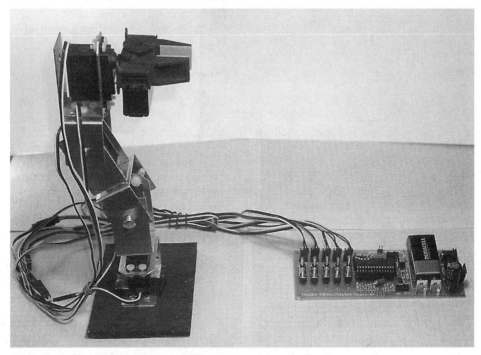

Figure 12.34 Finished robotic arm and five-servomotor controller.

```
'Slow-speed
'Manual control of five servomotors using 5 SPDT switches
'Microcontroller PIC 16F873

adcon1 = 7                 'Set port a to digital I/O

'Declare variables

b0 var byte           'Use b0 as hold pulse width variable for servo 1
b1 var byte           'Use b1 to hold pulse width variable for servo 2
b2 var byte           'Use b2 to hold pulse width variable for servo 3
b3 var byte           'Use b3 to hold pulse width variable for servo 4
b4 var byte           'Use b4 to hold pulse width variable for servo 5
b6 var byte           'Variable for pause routine
b7 var word           'Variable for pause routine
s1 var byte           'Unassigned delay variable
s2 var byte           'Assigned delay variable

'Initialize servomotor variables
b0 = 150              'Start up position servo 1
b1 = 150              'Start up position servo 2
b2 = 150              'Start up position servo 3
b3 = 150              'Start up position servo 4
b4 = 150              'Start up position servo 5

s2 = 4               'Delay variable

start:

'Output servomotor position

portb = 0           'Prevents potential signal inversion on reset
   pulsout portb.7, b0  'Send current servo 1 position out
   pulsout portb.6, b1  'Send current servo 2 position out
   pulsout portb.5, b2  'Send current servo 3 position out
   pulsout portb.4, b3  'Send current servo 4 position out
   pulsout portb.3, b4  'Send current servo 5 position out

'Routine to adjust pause value (nom 18) to generate approx 50 Hz update

b7 = b0 + b1 + b2 + b3 + b4
b6 = b7/100
b7 = 15 - b6
pause b7

'Check for switch closures
   if portc.3 = 0 then left5    'Is sw1 left active?
   if portc.2 = 0 then right5   'Is sw1 right active?
   if portc.1 = 0 then left4    'Is sw2 left active?
```

```
    if portc.0 = 0 then right4     'Is sw2 right active?
    if porta.5 = 0 then left3      'Is sw3 left active?
    if porta.4 = 0 then right3     'Is sw3 right active?
    if porta.3 = 0 then left2      'Is sw4 left active?
    if porta.2 = 0 then right2     'Is sw4 right active?
    if porta.1 = 0 then left1      'Is sw5 left active?
    if porta.0 = 0 then right1     'Is sw5 right active?

    goto start

'Routines for servomotor 1
left1:
    s1 = s1 + 1
    if s1 = s2 then
    b0 = b0 + 1                'Increase the pulse width
    s1 = 0
    endif
    if b0 > 254 then max0       'Maximum 2.54 milliseconds
    goto start
right1:
    s1 = s1 + 1
    if s1 = s2 then
    b0 = b0 - 1                'Decrease the pulse width
    s1 = 0
    endif
    if b0 < 75 then min0        'Minimum .75 millisecond
    goto start
max0:
    b0 = 254                   'Cap max b1 at 2.54 milliseconds
    goto start
min0:
    b0 = 75                    'Cap min b1 at .75 millisecond
    goto start

'Routines for servomotor 2
left2:
    s1 = s1 + 1
    if s1 = s2 then
    b1 = b1 + 1                'Increase the pulse width
    s1 = 0
    endif
    if b1 > 254 then max1       'Maximum 2.54 milliseconds
    goto start
right2:
    s1 = s1 + 1
    if s1 = s2 then
    b1 = b1 - 1                'Decrease the pulse width
    s1 = 0
    endif
    if b1 < 75 then min1        'Minimum .75 millisecond
```

```
      goto start
max1:
  b1 = 254                    'Cap max b1 at 2.54 milliseconds
  goto start
min1:
  b1 = 75                     'Cap min b1 at .75 millisecond
  goto start


'Routines for servomotor 3
left3:
  s1 = s1 + 1
  if s1 = s2 then
  b2 = b2 + 1                 'Increase the pulse width
  s1 = 0
  endif
  if b2 > 254 then max2       'Maximum 2.54 milliseconds
  goto start
right3:
  s1 = s1 + 1
  if s1 = s2 then
  b2 = b2 - 1                 'Decrease the pulse width
  s1 = 0
  endif
  if b2 < 75 then min2        'Minimum .75 millisecond
  goto start
max2:
  b2 = 254                    'Cap max b2 at 2.54 milliseconds
  goto start
min2:
  b2 = 75                     'Cap min b2 at .75 millisecond
  goto start


'Routines for servomotor 4
left4:
  s1 = s1 + 1
  if s1 = s2 then
  b3 = b3 + 1                 'Increase the pulse width
  s1 = 0
  endif
  if b3 > 254 then max3       'Maximum 2.54 milliseconds
  goto start
right4:
  s1 = s1 + 1
  if s1 = s2 then
  b3 = b3 - 1                 'Decrease the pulse width
  s1 = 0
  endif
  if b3 < 75 then min3        'Minimum .75 millisecond
  goto start
max3:
```

```
    b3 = 254                        'Cap max b3 at 2.54 milliseconds
    goto start
min3:
    b3 = 75                         'Cap min b3 at .75 millisecond
    goto start


'Routines for servomotor 5
left5:
    s1 = s1 + 1
    if s1 = s2 then
    b4 = b4 + 1                     'Increase the pulse width
    s1 = 0
    endif
    if b4 > 254 then max4           'Maximum 2.54 milliseconds
    goto start
right5:
    s1 = s1 + 1
    if s1 = s2 then
    b4 = b4 - 1                     'Decrease the pulse width
    s1 = 0
    endif
    if b4 < 75 then min4            'Minimum .75 millisecond
    goto start
max4:
    b4 = 254                        'Cap max b4 at 2.54 milliseconds
    goto start
min4:
    b4 = 75                         'Cap min b4 at .75 millisecond
    goto start

end
```

In the above program variable S2 is assigned a value of 4. To increase the speed of the servomotor's movement, decrease this value. To slow down the servomotor movement, increase this value.

Increasing the Lifting Capacity of the Robotic Arm

Substituting the top two HS-322 servomotors connected to the gripper with two HS-85MG servomotors can increase the lifting capacity of the robotic arm. The HS-85MG servomotors are substantially smaller and lighter, while producing close to the same torque as the HS-322 servomotors. The downside is that the HS-85MG servomotors cost about 3 times the amount of the HS-322 servomotors. Do not try to substitute the HS-85BB servomotor for the HS-85MG. The HS-85BB uses plastic gears, which will strip pretty quickly. The HS-85MG incorporates metal gears that last.

To use the HS-85MG servomotors in the robotic arm, substitute the top HS-322 bracket for an HS-85MG bracket. In addition you need to order the servomotor gripper that has been modified to use an HS-85MG servomotor.

Adding a Robotic Arm Base

The weakest link in the robotic arm, as it stands right now, is the base servomotor. The bearing in the bottom servomotor is subjected to all the stress and weight of the entire arm as it turns and lifts any object. We can greatly improve upon this situation by adding a second bearing that removes most of the stress on the servomotor's small bearing. To incorporate this second bearing, we need to build a small base.

I tried a number of designs. The one that I feel works best is made primarily from $3/4$-in-thick hardwood. The following drawings show the five pieces needed to make the base. Figures 12.35 and 12.36 show the wood blocks needed for mounting the base servomotor. Figures 12.37 and 12.38 show the sides for the base. Figure 12.39 is a metal baseplate. The two servomotor blocks are mounted to the baseplate, using wood screws through the bottom. The servomotor is mounted to the wood blocks (see Fig. 12.40). Next the side pieces are mounted to the wood block (see Fig. 12.41). We need a 0.40-in, "length of 1"-in-diameter wood dowel. To this piece of wood we center and attach a round servomotor horn, using two small wood screws (see Fig. 12.42). The top of the servomotor horn should be sanded flat to remove the small lip around the center.

The wood dowel is fitted onto the base servomotor (see Fig. 12.43). Next the 3-in-square bearing is placed on the sides to ensure everything lines up properly. The wood dowel should be centered in the bearing (see Fig. 12.44). Mount the bearing to the sides, using four wood screws.

A top plate for the 3-in-square bearing is shown in Fig. 12.45. This plate is mounted to the bearing using four 6-32 plastic machine screws and nut.

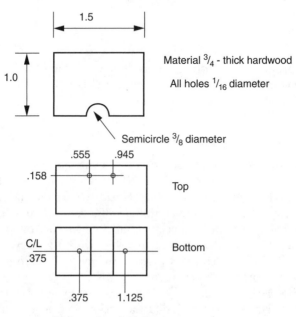

All dimensions in inches

Figure 12.35 Servomotor block A.

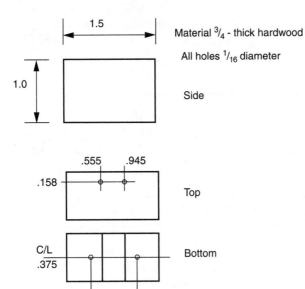

Material $^3/_4$ - thick hardwood

All holes $^1/_{16}$ diameter

Side

Top

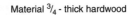

Bottom

All dimensions in inches

Figure 12.36 Servomotor block B.

Material $^3/_4$ - thick hardwood

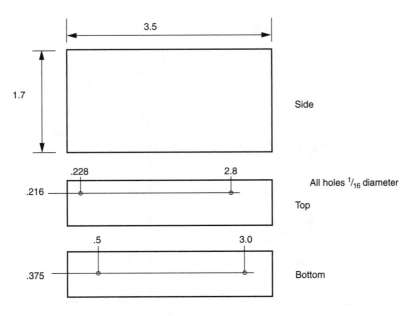

Side

All holes $^1/_{16}$ diameter

Top

Bottom

All dimensions in inches

Figure 12.37 Side block A.

Material $^3/_4$ - thick hardwood

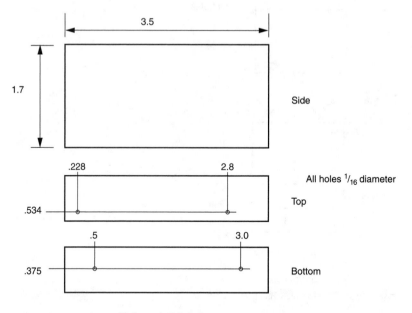

All holes $^1/_{16}$ diameter

All dimensions in inches

Figure 12.38 Side block B.

1/8 - 3/16 aluminum or CRS

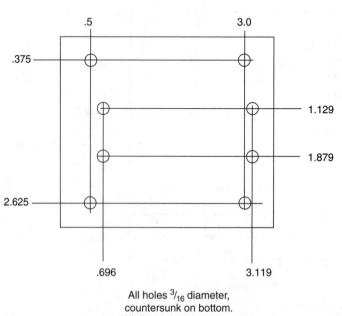

All holes $^3/_{16}$ diameter,
countersunk on bottom.
All dimensions in inches.

Figure 12.39 Baseplate.

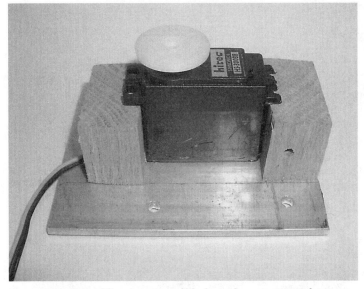

Figure 12.40   Assembling servomotor blocks and servomotor to base.

Figure 12.41   Attaching sides to base.

Figure 12.42 A 1-in × 0.40-in wood dowel with round servomotor horn.

Figure 12.43 Attaching a servomotor horn to servomotor base.

When the top plate is secured to the bearing, the top of the wood dowel should be right underneath the top plate. Place the bottom servomotor bracket of the robotic arm on top of the top plate. Secure the servomotor bracket (and top plate) to the underlying dowel through the four center holes in the top bearing plate (see Fig. 12.46).

The top section of the robotic arm is fitted to the base servomotor bracket. The finished robotic arm is shown in Figs. 12.47 and 12.48. In the picture note the use of the smaller HS-85MG servomotors connected to the gripper.

Figure 12.44   Attach 3-in-square bearing to base, check for alignment.

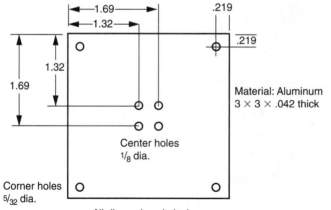

All dimensions in inches

Figure 12.45   Top bearing plate.

Figure 12.46 Attach top bearing plate, servomotor bracket to 3-in-square bearing.

Figure 12.47 Finished robotic arm with base (right side).

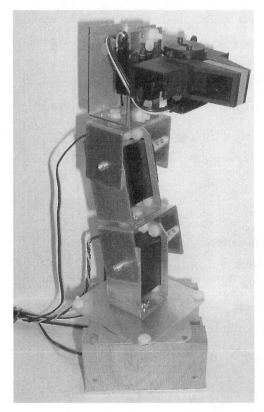

Figure 12.48 Finished robotic arm with base (left side).

Parts List
Robotic arm

(3) HiTec servomotors (HS-322HD)

(2) HiTec servomotors (HS-475HB)

(5) Servomotor bracket assemblies

(1) Servomotor gripper assembly

(1) Base board

(5) 12- or 24-in servomotor extensions

Base

Servomotor blocks A and B

Baseplate

Base sides A and B

3-in-square bearing

Top bearing plate

1-in-diameter $\times$ 0.40-in wood dowel

Plastic machine screws and nuts, wood screws

Available from Images SI Inc. (see Suppliers at end of book).

Four-servomotor controller

(1) PIC 16F84

(1) 4-MHz Xtal

(2) 22-pF capacitors

(4) SPDT PC-mounted switches with center-off position

(8) 10-kΩ, $\frac{1}{4}$-W resistors

(1) 4.7-kΩ, $\frac{1}{4}$-W resistor

(1) 0.1-μF, 50-V capacitor

5-V power supply (regulated)

Kit available from Images SI Inc. (see Suppliers).

Five-servomotor controller

(1) PIC 16F873

(1) 4-MHz Xtal

(2) 22-pF capacitors

(5) SPDT PC-mounted switches with center-off position

(11) 10-kΩ, $\frac{1}{4}$-W resistors

(1) 4.7-kΩ resistor

(1) 0.1-μF, 50-V capacitor

5-V dc power supply

Kit available from Images SI Inc. (see Suppliers).

13

Bipedal Walker Robot

In this chapter we will construct and program a bipedal walking robot (see Fig. 13.1). Bipedal robots more closely resemble humans because they use two legs to walk. Bipedalism is a necessary step to creating advanced robots that can work and function in human environments. The heart and mind of this robot are the 16F84 microcontroller. The microcontroller will be programmed using the PicBasic (or PicBasic Pro) compiler. Muscle for motion is generated using a series of eight HS-322 servomotors, four servomotors for each leg.

I have not taken any shortcuts in building this bipedal robot, meaning this robot is a true bipedal walker robot. This criterion demands that the robot balance itself on one leg in order to lift the other leg to initiate walking. This action is accomplished using independent ankle, knee, and hip movements. This bipedal robot does not have oversized feet or footpads. This eliminates the type of low-technology tilting bipedal walker that uses oversized feet to keep the robot from tipping over when movement proceeds from one leg to the other. You may have seen this type of "big-foot" walker; the older units have a motor-activated cam that rises and moves one leg after the other. Lately I've seen servomotor-powered tilting big-foots on the loose.

To see a movie of this bipedal robot walking, go to the Internet to the following page: www.imagesco.com/catalog/biped/walker.html.

My design calls for using four servomotors in each leg (see Fig. 13.1). The initial walking gait programmed into the robot resembled that of the flamingo bird. This particular bird has a reverse knee joint. If that bird doesn't bring a clear enough picture to mind, perhaps the Imperial walker from the original Star Wars film(s) will suffice.

Nature has evolved a three-jointed leg for most walking animals. Although it may appear that our robotic leg has four joints, because it has four servomotors, it is essentially a three-jointed leg. The reason is that our first and second servomotors, starting from the bottom of the leg, form a two-directional ankle. It is important that the ankle can tilt the foot, left to right as well as

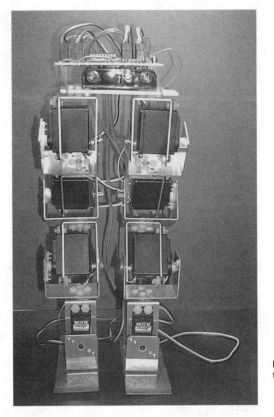

Figure 13.1 Bipedal robot ready to walk.

forward and backward. Humans have two-directional ankles; this requires two servomotors to replicate in our leg.

So the third servomotor is considered the knee joint, and the forth servomotor the hip joint.

A Question of Balance?

When we walk, we receive constant feedback from our leg muscles and feet such as stretch, tension, and load, in addition to having tilt and balance information present from our inner ear. Remove this physical feedback information and remove any visual clues, and it becomes much harder to walk. Imagine how much harder, if not impossible, it would be to learn how to walk without sensory feedback.

This lack of feedback is a dilemma for robotics. It is possible to program a bipedal walker robot to walk without feedback and a sense of balance. To do so, exact position control and movements are measured for each leg servomotor action, each action sequence is programmed into the microcontroller, the program is initiated, and the sequence repeated to achieve a walking gait.

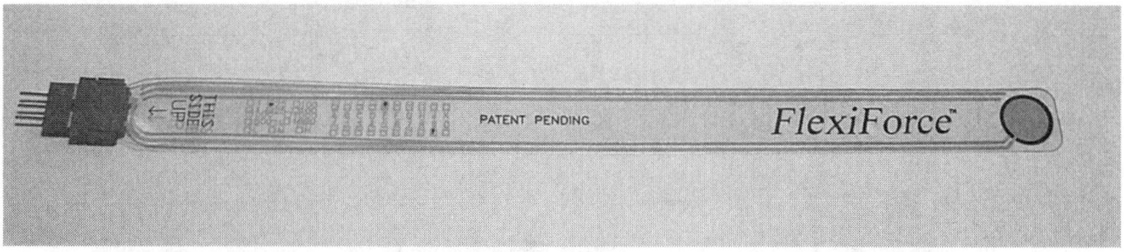

Figure 13.2 FlexiForce pressure sensor.

This brute-force programming works, but it is not adaptive. If any weight on the robot shifts (battery pack moves) or if you have the robot carry a weight, anything that changes the robot's center of gravity, then the program will need to be adjusted. A little sensory feedback may help the robot walk and be more adaptive.

A Little Feedback

Feedback comes in many forms. The sensor I would incorporate into this robot is a pressure sensor. I will be placing a pressure sensor on the base of each footpad. The sensor could tell the microcontroller when there is no pressure (weight) on a foot. This could be used to adaptively tilt the robot until there is no weight on the opposite footpad.

The sensor is a FlexiForce pressure sensor (see Fig. 13.2). (FlexiForce is a trademark of Tekscan, Inc.) This particular sensor is made to detect pressure from 0 to 1 lb. Although the final weight of the robot may be slightly more the sensor top weight, I feel it's a better (more sensitive) choice than taking the next sensor that measures pressure between 0 and 25 lb.

The pressure sensor is a variable-resistor type. As pressure increases, its resistance drops. Since we are using the sensor to determine when there is zero weight on a leg, we don't need to perform an A/D conversion to read varying pressure (weight). Instead we can use an op-amp and comparator. The op-amp converts the resistance change in the sensor to an electric change. The comparator is set to trigger on zero weight. The output of the comparator can be read by the microcontroller as a simple high-low signal.

This bipedal robot does not use any feedback, so it is not adaptive to shifting weight loads. I have provided this feedback information in case you wish to advance this basic bipedal walker on your own.

Servomotors

This bipedal walker utilizes common inexpensive HiTec HS-322HD 42-oz torque servomotors. Other more powerful servomotors are available, such as the HS-425 and HS-475, and they will increase the weight-carrying capacity of the robot. However, these more powerful servomotors also require greater electric current. So the battery pack will need to be increased proportionally. The robot, as it stands, is capable of carrying its own 6-V battery pack and circuitry.

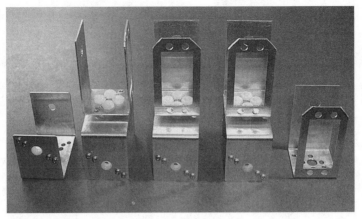

Figure 13.3 Servomotor brackets needed for one leg.

Servomotor Brackets

This robot uses the same servomotor brackets as outlined in Chap. 12. That information will not be repeated here. In Fig. 13.3 the brackets needed for one robotic leg are shown. You need two such sets of servomotor brackets, eight in all, to build this bipedal robot. The servomotor horns used on these servomotor brackets are included with all the compatible HiTec servomotors, such as HS-322, HS-425, HS-475, and HS-35645. These brackets may also be used with similar-size Futaba servomotors, but you may have to purchase the horns separately.

Footpads

The footpads for the robot are shown in Figs. 13.4 and 13.5. I glued rubber gasket material to the bottom of the plastic footpad to make the pad nonskid.

The footpads provide a larger surface area that makes it easier for the biped to balance and walk. They are attached to the bottom U bracket of the bottom servomotor. I arbitrarily chose to make the footpad size 1.5 in wide × 4 in long. I cut out this size rectangle from $1/_4$-in-thick acrylic plastic. The location of the servomotor bracket on the feet is shown in Fig. 13.4. You will notice the bracket is not centered on the plastic foot; it is located at one side toward one end (considered the back). Drill four $1/_8$-in-diameter holes in the plastic that line up with the four holes on the U bracket. Each drilled hole must be countersunk on the bottom of the foot, so that the machine screw head will not protrude from the bottom of the foot; see side view and close-up of Fig. 13.4 and finished footpad in Fig. 13.5. This will allow the foot to lie flat against the floor.

On the prototype the corners of the footpads are square (see Fig. 13.5). I plan to round the corners of the footpads, so they will be less likely to catch on something and trip the robot when walking. The footpads are attached to the U bracket using four 4-40 machine screws, nuts, and lockwashers.

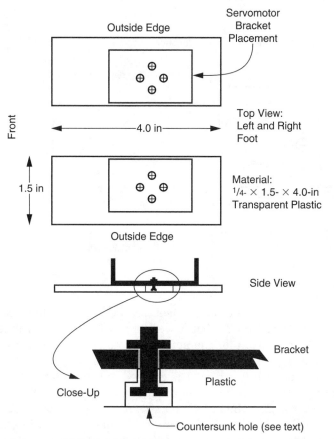

Outside Edge

Servomotor
Bracket
Placement

Front

◄——————4.0 in——————►

Top View:
Left and Right
Foot

1.5 in

Material:
$^{1}/_{4}$- × 1.5- × 4.0-in
Transparent Plastic

Outside Edge

Side View

Bracket

Plastic

Close-Up

Countersunk hole (see text)

Figure 13.4 Diagram of footpad.

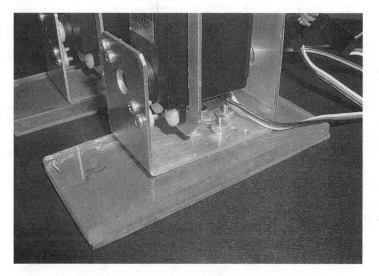

Figure 13.5 Picture of footpad.

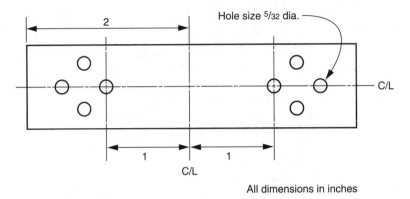

Material $^1/_8 \times 1 \times 4$ aluminum

Figure 13.6   Aluminum hip bar.

The bottom of the acrylic plastic feet can be slippery, depending upon the surface material the bipedal robot is walking on. I glued soft rubber sheet gasket material to the bottom of the acrylic feet to create a nonskid bottom surface for the feet. If just the front and back of the gasket material are glued to the plastic foot, a small flat pocket is created in the center section of the foot. This flat pocket is ideal for locating a flat sensor that could be slid in between the gasket material and the acrylic plastic. Although we will not be using any flat sensor in this robot, it could become a future modification, and you may want to leave this option open when gluing the gasket material to the footpad.

I have found this robot biped walks and balances so easily that I believe it's possible to reduce the size of the footpads or remove them entirely. This idea is open for future experimentation.

The hip bar that connects the top servomotor brackets of both legs is shown in Fig. 13.6. The base material is $^1/_8$-in-thick aluminum bar 1 in wide $\times$ 4 in long. Mark a centerline (C/L) across the width and the length, as shown in Fig. 13.6. From the width C/L mark another line 1 in away from the C/L on each side. Next use the base of the servomotor bracket to mark the four mounting holes. Align the bracket on the left side so that an "X" from the drawn centerlines is centered in the rightmost hole. Mark the four holes with a pencil. Align the bracket on the right side so that an "X" from the drawn centerlines is centered in the leftmost hole. Mark the four holes with a pencil.

Punch the center of each hole with a hammer and punch. Drill the punch holes with a $^5/_{32}$-in drill. Clean each hole to remove any burrs with a file or deburring tool.

Assembly

When you assemble the servomotors to the servomotor brackets, center each servomotor before attaching the servomotor shaft to the horn-bracket assem-

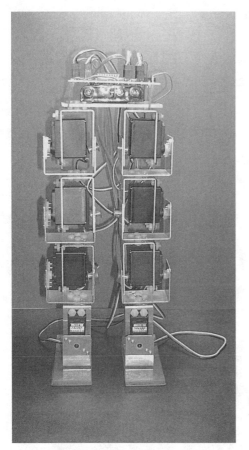

Figure 13.7 Bipedal robot with all servomotors centered.

bly. The walking program expects the servomotors to be aligned in this way. If a centering servomotor signal is sent to all eight servomotors, the robot walker will appear as shown in Fig. 13.7. This is not the start position of the walking program.

Schematic

Figure 13.8 is the schematic of our bipedal walker robot. To achieve maximum torque from the servomotors, I needed to run them at 6 V. To run the PIC 16F84 at close to 5 V, I incorporated a 1N4007 diode. The average voltage drop across a silicon diode is 0.7 V. So at peak power from the batteries (under load) the microcontroller will receive about 5.3 V, which is within the voltage range for this microcontroller. A photograph of the prototype circuit is shown in Fig. 13.9. The battery pack I used is below the circuit board. It holds four AA batteries. I used a small piece of Velcro to secure the battery pack to the hip bar. I secure the circuit board by using two small elastic bands (see Fig. 13.10).

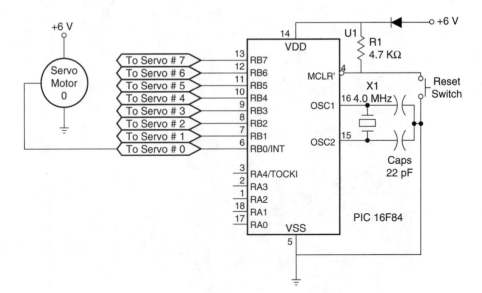

Figure 13.8 Bipedal robot schematic.

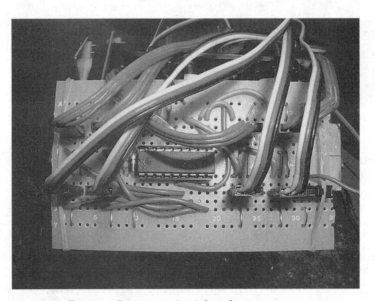

Figure 13.9 Top view of prototype circuit board.

The four AA battery, 6-V power supply only lasts a short time. The bipedal robot appears to be able to lift more weight than I placed on it, so you may be able to add a second 6-V power supply and increase the untethered walking time. In any case I only use the battery pack for demonstrations. For most development

Figure 13.10   Side view of circuit board and battery pack attached to robot.

work you may want to build an external regulated power supply for the biped, as I have, and tether the power supply to the robot. Keep the unused battery pack on the robot, so you will not have to compensate for the additional weight when demonstrating the robot's walking ability using the battery pack.

Program

When the robot is assembled, you may have to adjust the program slightly. There will be slight variances in your servomotor positions as compared to my prototype due to small variances in the construction. You only need to add or remove one line in the entire program to make adjustments, and the line is:

```
goto hold
```

The `hold` subroutine keeps the servomotors locked in their last position. The robot stays frozen, giving you plenty of time to look over its position.

This is the procedure for using that one line and adjusting the program. You place that line after each robotic movement. Check the position, adjust the movement if necessary, check again, and adjust if necessary until the position is perfect. Movement is adjusted by varying the Y1 and Y2 numbers in each movement. I cannot imagine the variance being more that ±5 points off what the program is showing.

There are 15 movements to check. I would advise letting the robot step through each movement; you will see if there is a problem. The robot may either trip on its feet or lose its balance. If that happens, you know you have to adjust that movement. But you must work it through movement by move-

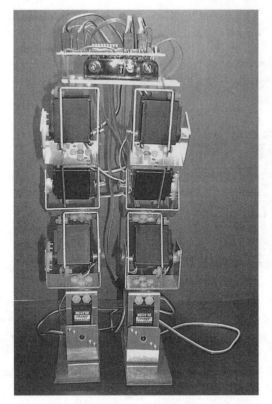

Figure 13.11   Front view of robot.

ment. If you just try to let the walker walk, it will be hard for you to determine which movement (if any) is causing a problem.

The first thing to check is the start position of the robot. Write the `goto hold` line right after the command `Gosub servoout`. The robot should be level, standing in a position shown in Figs. 13.11 and 13.12.

If adjustments are necessary, you need to make them in the "initialize variables" section. Once you are satisfied, remove the `goto hold` line you wrote in the program. Place the `goto hold` line at the end of the "First movement." Check position, adjust if necessary, then move the `goto hold` line to the end of the "Second movement." Continue in this manner until all movements have been checked.

The way the program is written, the robot will take three steps and then stop. You can change the range of B(10) to increase or decrease the amount of steps taken.

Subroutines M1, M2, and M3

The subroutines M1, M2, and M3 are delay routines. These routines slow the servomotor movement, so the movement is smooth. Without these routines the

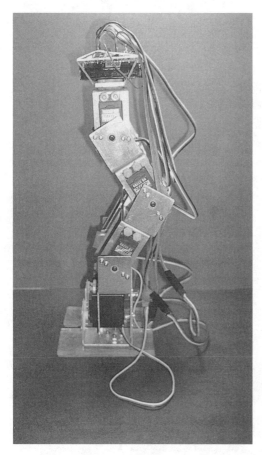

Figure 13.12 Side view of robot.

servomotors would jerk into position so quickly that the motion would topple the robot. The reason for three routines is that I want to affect two independent servomotor motions at the same time. The numbers controlling the servomotor positions could be both (1) decreasing (M1 –,–) and increasing (M2 +,+) and (2) increasing and decreasing (M3 +,–). Hence we need the three subroutines to handle the motion.

```
'Bipedal walker program

'Declare variables

x1 var byte
x2 var byte
y1 var byte
y2 var byte
lp var byte

'Declare array
```

```
b var byte[12]

'Initalize array variables

b(0)  = 148          'Right ankle (vertical)
b(1)  = 121          'Right ankle (horiz.)
b(2)  = 204          'Right knee
b(3)  = 126          'Right hip
b(4)  = 150          'Left ankle (vertical)
b(5)  = 178          'Left ankle (horiz.)
b(6)  = 101          'Left knee
b(7)  = 180          'Left hip
b(8)  = 0            'Counter
b(9)  = 0            'Counter
b(10) = 0            'Counter
b(11) = 0            'Dummy value

start:

'Holding loop that holds upright position 3 seconds before moving

b(8) = b(8) + 1

gosub servoout

if b(8) < 180 then goto start

b(8) = 0                       'Reset loop counter

'_____


for b(10) = 1 to 3             'Take 3 steps forward

'_____


'Leg movements for one whole step
'_____

'First movement
   x1 = 0                      'Servomotor 0
   x2 = 4                      'Servomotor 4
   y1 = 129                    'Tilt right ankle (horiz.)
   y2 = 135                    'Tilt left ankle (horiz.)
   lp = 106                    'Loop counter
   gosub m1
'_____


'Second movement
   x1 = 5                      'Servomotor 5
   x2 = 6                      'Servomotor 6
```

```
    y1 = 2              'Left ankle (vert.)
    y2 = 70             'Left knee
    lp = 140            'Loop counter
    gosub m3
'_____


'Third movement
    x1 = 4              'Servomotor 4
    x2 = 7              'Servomotor 7
    y1 = 150            'Left ankle
    y2 = 160            'Left hip
    lp = 75             'Loop counter
    gosub m3
'_____


'Fourth movement
    x1 = 1              'Servomotor 1
    x2 = 5              'Servomotor 5
    y1 = 132            'Straighten ankle
    y2 = 200            'Straighten ankle
    lp = 90             'Loop counter
    gosub m3
'_____


'Fifth movement
    x1 = 6              'Servomotor 6
    x2 = 0              'Servomotor 0
    y1 = 85             'Left knee
    y2 = 140            'Straighten ankle
    lp = 96             'Loop counter
    gosub m2
'_____


'Sixth movement
    x1 = 0              'Servomotor 0
    x2 = 4              'Servomotor 4
    y1 = 167            'Tilt ankle left
    y2 = 169            'Tilt ankle left
    lp = 80             'Loop counter
    gosub m2
'_____


'Seventh movement
    x1 = 6              'Servomotor 6
    x2 = 5              'Servomotor 5
    y1 = 101            'Tilt knee
    y2 = 178            'Tilt ankle
    lp = 95             'Loop counter
    gosub m3
'_____
```

```
'Eighth movement
    x1 = 2              'Servomotor 2
    x2 = 1              'Servomotor 1
    y1 = 244            'Right ankle
    y2 = 104            'Right knee
    lp = 140            'Loop counter
    gosub m3
'_____


'Ninth movement
    x1 = 3              'Servomotor 3
    x2 = 7              'Servomotor 7
    y1 = 140            'Right hip
    y2 = 180            'Left hip
    lp = 80             'Loop counter
    gosub m2
'_____


'Tenth movement
    x1 = 1              'Servomotor 1
    x2 = 2              'Servomotor 2
    y1 = 121            'Right ankle
    y2 = 204            'Right knee
    lp = 150            'Loop counter
    gosub m3
'_____


'Eleventh movement
    x1 = 0              'Servomotor 0
    x2 = 4              'Servomotor 4
    y1 = 129            'Straighten right ankle
    y2 = 135            'Straighten left ankle
    lp = 150            'Loop counter
    gosub m1
'_____


'Twelfth movement
    x1 = 5              'Servomotor 5
    x2 = 6              'Servomotor 6
    y1 = 217            'Left ankle
    y2 = 70             'Left knee
    lp = 144            'Loop counter
    gosub m3
'_____


'Thirteenth movement
    x1 = 4              'Servomotor 4
    x2 = 3              'Servomotor 3
    y1 = 133            'Left ankle
    y2 = 126            'Right hip
```

```
    lp = 66                 'Loop counter
    gosub m1
'_____


'Fourteenth movement
    x1 = 6                  'Servomotor 6
    x2 = 5                  'Servomotor 5
    y1 = 101                'Left knee
    y2 = 178                'Left ankle
    lp = 144                'Loop counter
    gosub m3
'_____


'Fifteenth movement
    x1 = 0                  'Servomotor 0
    x2 = 4                  'Servomotor 4
    y1 = 148                'Left knee
    y2 = 150                'Left ankle
    lp = 115                'Loop counter
    gosub m2
'_____

    next b(10)              'Next step
'_____


hold:                       'Hold position
    gosub servoout
    goto hold
    '_____


servoout:

    'Output servomotor position(s)

    portb = 0               'Prevent signal inversion

'Right leg
    pulsout portb.0, b(0)   'Send current servo 1 position out
    pulsout portb.1, b(1)   'Send current servo 2 position out
    pulsout portb.2, b(2)   'Send current servo 3 position out
    pulsout portb.3, b(3)   'Send current servo 4 position out

'Left leg
    pulsout portb.4, b(4)   'Send current servo 5 position out
    pulsout portb.5, b(5)   'Send current servo 6 position out
    pulsout portb.6, b(6)   'Send current servo 7 position out
    pulsout portb.7, b(7)   'Send current servo 8 position out

    pause 5                 '5 millisecond delay to generate 50 Hz signal
    return                  'To servomotors
    '_____
```

```
m1:                                    `(negative increment(s) -,-)
    b(8) = b(8) + 1
        if b(9) = 2 then m12
    b(9) = b(9) + 1
    goto m13

    m12:
        b(x1) = b(x1) - 1
        b(x2) = b(x2) - 1
        b(9) = 0

    m13:
        if b(x1) < y1 then
            b(x1) = y1
        endif
        if b(x2) < y2 then
            b(x2) = y2
        endif
    gosub servoout
    if b(8) < lp then m1
    b(x1) = y1
    b(x2) = y2
    b(8) = 0
    b(9) = 0
    return
        `_____

m2:                                    `(positive increment(s) +,+)
    b(8) = b(8) + 1
        if b(9) = 2 then m22
    b(9) = b(9) + 1
    goto m23

    m22:
        b(x1) = b(x1) + 1
        b(x2) = b(x2) + 1
        b(9) = 0
    m23:
        if b(x1) > y1 then
            b(x1) = y1
        endif
        if b(x2) > y2 then
            b(x2) = y2
        endif
    gosub servoout
    if b(8) < lp then m2
    b(x1) = y1
    b(x2) = y2
    b(8) = 0
    b(9) = 0
```

```
        return
        '_____


m3:                                  '(positive - negative increment +,-)
    b(8) = b(8) + 1
        if b(9) = 2 then m32
    b(9) = b(9) + 1
    goto m33
    m32:
        b(x1) = b(x1) + 1
        b(x2) = b(x2) - 1
        b(9) = 0
    m33:
        if b(x1) > y1 then
            b(x1) = y1

        endif

        if b(x2) < y2 then
            b(x2) = y2

        endif

    gosub servoout
    if b(8) < lp then m3
    b(x1) = y1
    b(x2) = y2
    b(8) = 0
    b(9) = 0
    return
```

Going Further

There are many areas for improvement. One of the simplest tasks you can perform is to reduce the loop counter (LP) variable in each movement. I exaggerated this number to ensure that the servomotors got to their proper position.

The walking gait used in this robot was the first one I developed. I am sure there is much room for improvement for anyone who wants to take the time and develop one. In addition, you can try to program completely different walking gaits. Right now the robot used two reverse knee joints. I looked at the robot stance using one reversed knee and one forward knee. It appears to have better been balanced than the current two reverse knee biped stance. In the future I may try to develop a gait using a forward and reverse knee stance. This would most definitely be a robotic gait, since I don't believe there is any animal that uses both a reverse and a forward knee leg for locomotion. This is another area you may want to work on.

Turning right and left

As the bipedal walker stands, it can only walk forward. While I was developing the walking program, I happened across an interesting accident. On certain occasions the robot would pivot to the left or to the right. I plan on developing this "accident" to see if I can use it to turn the robot to the left and right. If you want to attempt this, here are the basic instructions. To make the robot pivot, first raise one leg. On the raised leg tilt the horizontal ankle servomotor slightly, and then tilt the vertical ankle servomotor up slightly. Next place the weight back down on the raised leg; the robot will pivot as the weight shifts. Turning the robot in this manner must be accomplished incrementally. Try to turn it too much at one time, and the robot will topple.

Finally with a little work, you should be able to make the robot walk backward.

I mentioned adaptive walking and balance control. We used all eight pins of port B on the PIC 16F84, but the PIC 16F84 still has five unused pins assigned to port A. These pins could be used for programming options such as adaptive balance control, or a run/walk switch, perhaps a forward/backward switch, or even a turn left/right sensor.

So you see, we have only scratched the surface of playing with this biped walker. There is much one can do and learn from this project.

Parts List

(8) HiTec servomotors (or similar-size and -torque servomotor) (HS-322)

(8) Servomotor brackets

(2) $1/4$-in $\times$ 1.5-in $\times$ 4-in acrylic plastic

(1) PIC 16F84 (4-MHz)

(1) 4.0-MHz Xtal

(2) 22-pF capacitors

(1) Diode (1N4007)

(1) 6-V AA battery holder (flat)

(8) Three-position headers (for connecting servomotors to circuit)

Plastic screws and nuts, Velcro, prototyping breadboard, elastic bands

14

Color Robotic Vision System

The robot we will build in this chapter will be capable of looking at and seeing an object (or target) and following that object. If the object gets too close to the robot, the robot will back away from it. Choose a bright target that will have good contrast with the background environment. Colored objects are fine; a bright red or yellow ball works fine. For my tests I used a 2.5-in square of orange construction paper taped to a stiff wire.

One of the most difficult areas in robotics today is the creation of an artificial vision system for a robot to see. Teaching a robot to see is not simply a case of connecting a video camera to a computer. The electronic representation of an image created by a video camera must be presented in a way a computer can "look at" and interpret ("see") the image.

To gain an understanding of the processes involved, let's examine how a computer might look at a simple black-and-white image. We must first define the resolution of the picture image in pixels. For our discussion let's assume a low-resolution picture of 80×143 pixels. At that resolution our computer must look at 11,440 pixels ($80 \times 143 = 11,440$). Each pixel of the picture can be any tonality of gray between pure black and pure white. We now have to determine how many different shades our computer can differentiate between pure black and pure white. If we wanted each pixel to be represented by a single byte (8-bit number), there would be 256 shades of gray, including pure white (0) and black (255).

The computer would look at each pixel and assign a number between 0 and 255 depending upon its tonality (grayness). After the computer assigned numbers to each 11,440 pixels, it transformed the basic image into a numbered representation it needs to "look" at the image.

The software that looks at an image and interprets visual features is appropriately called *image processing*. Robotists over the years have gleaned some techniques for helping computers to see. One technique is called *edge detection*. Here the computer looks through the image. You program the computer

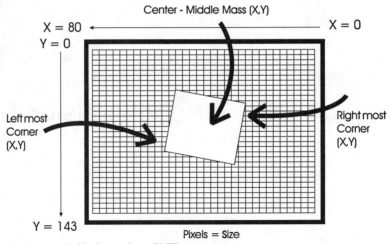

Figure 14.1   Field of view from CMU camera.

so that any pixel dark enough to be a 200 number or larger is a possible edge, and you change that pixel to pure black, number 256. Any number less than 200 probably isn't an edge, and the computer changes those pixels to pure white, number 1. What is left is a simplified representation of the picture that can be more easily analyzed.

The same process used for edge detection may also be used to detect particular colors in an image or the contrast between colors. Once an object has been detected, through contrast, color, or edge detection, the processing software can assign location parameters to the object within the image and field of view (FOV) of the camera.

In Fig. 14.1 we have a representation of the FOV from the CMU camera we will be using and a few of the image processing parameters available. Once an object is detected by the camera, we can read these image processing parameters in real time from the serial communication port of the camera. We use these parameters to track an object in the camera's image space and to move our robot accordingly.

CMU Camera

The CMU camera (see Fig. 14.2) was developed at Carnegie Mellon University (CMU). The CMU camera uses an SX28 microcontroller interfaced to an Omnivision OV6620 CMOS camera chip. The SX28 microcontroller does much of the image processing for us. We communicate with the camera via a standard RS-232 or TTL serial port.

A few of the CMU camera features are as follows:

Tracks user-defined color objects at 17 frames per second

Finds the center of the object

Figure 14.2 Front CMU camera.

Gathers mean color and variance data

Resolution of 80 × 143 pixels

Serial communication at 115,200, 38,400, 19,200, and 9600 Bd

Demo mode that automatically locks onto and drives a servomotor to track an object

Serial Communication

As stated, we communicate to the CMU camera via a serial interface. We will create a serial communication link between the CMU camera and both a personal computer (PC) and the PIC microcontroller. We will first look at the PC communication to the CMU camera.

Figure 14.3 is a simple Windows 98 program. It allows you to test the CMU camera and the serial communication link (port number and baud rate). You can adjust the PC's baud rate and serial port through drop-down menu items. This program may be downloaded without cost from this website: http://www.cmucam.com.

Before you start the Windows program, you need to set up the CMU camera's baud rate. Figure 14.4 shows the back of the CMU camera, where the male header is located to place various jumpers. The baud rate is selected using jumper 2 and jumper 3 on the back of the CMU camera:

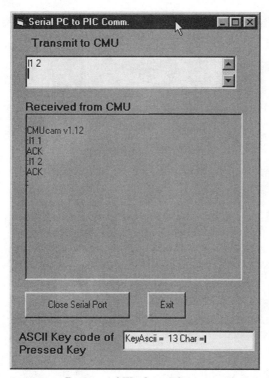

Figure 14.3 Basic serial Windows PC communication program.

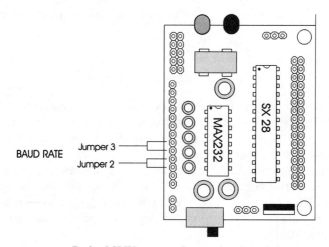

Figure 14.4 Back of CMU camera showing baud rate jumpers.

Baud rate	Jumper 2	Jumper 3
115,200	Open	Open
38,400	Set	Open
19,200	Open	Set
9,600	Set	Set

For PC communication I recommend using the 115,200-Bd rate. Use this baud rate because once you have your simple communication up and running, you can switch over to a more sophisticated Windows program to evaluate the CMU camera parameters.

With the baud rate set, connect the serial cable to the CMU camera (see Fig. 14.5). Connect a DB-9 pin serial cable from the PC to the camera.

Start the program. Set the program's baud rate to match the CMU camera's baud rate. Set the serial port to the one you connected to the CMU camera. If your computer has multiple serial ports, you may have to try different COMM ports to find out which one is connected to the camera.

To test a port, set the serial port to COMM1. Turn on the CMU camera. The following message should be displayed when the camera is turned on:

```
cmucam V1.12
:
```

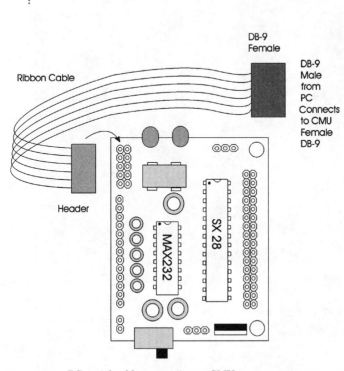

Figure 14.5 PC serial cable connection to CMU camera.

If you do not see this message, turn off the camera, set the serial port to COMM2, and test again. Continue in this manner until you find the right COMM port. If you don't see the message with any of the COMM ports on your computer, you may have the baud rate set improperly—double-check. Once you see the message, you begin to communicate with the CMU camera.

To turn on the camera's green LED, enter the command l1 1 and hit Return. To turn off the green LED, enter the command l1 2 and hit Return.

Once you have the communication link working, you are finished with the first program, and it is time to move onto the main VB application program.

VB Application Program

The VB application program is included on the CD-ROM with the CMU camera. The application allows you to see how the CMU camera images different scenes or targets. The current VB application isn't stable; however, by the time this book goes to press, a newer, (hopefully) more stable application program will be available. The simple application we used before provided the correct port number that you will need to allow this program to function properly. The baud rate used on this program is fixed at 115,200. So make sure the CMU camera is set at 115,200 Bd.

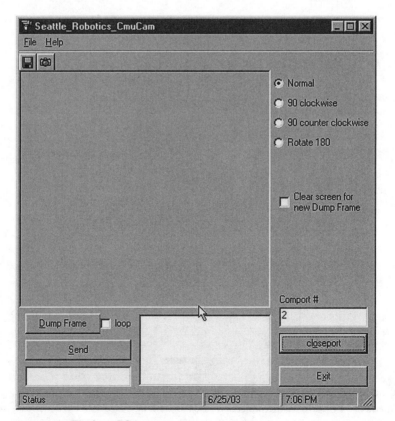

Figure 14.6 Windows PC program.

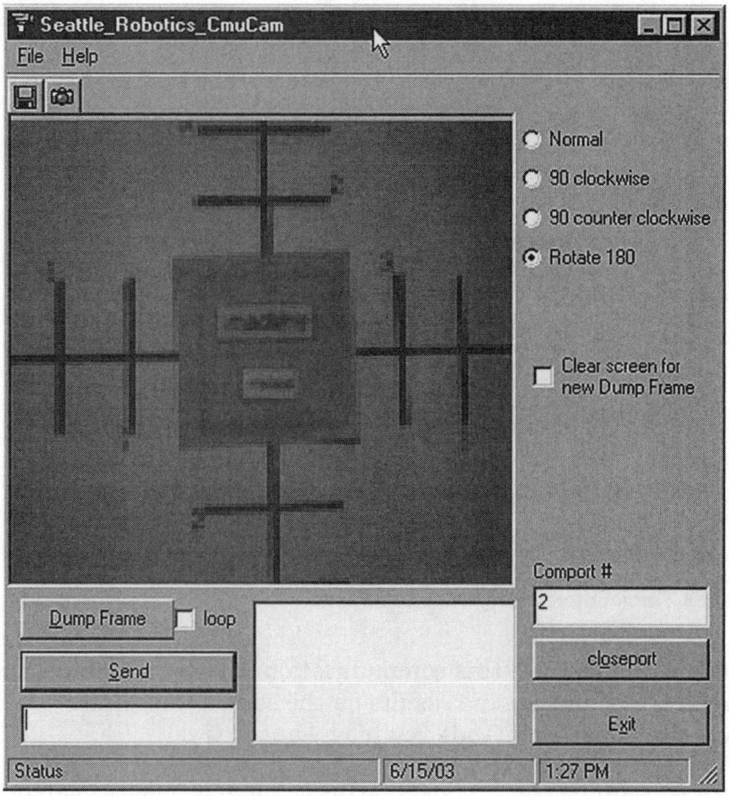

Figure 14.7 Windows PC program showing frame dump.

To view the image properly from the camera, hit the 180° option (see Fig. 14.6). Select the proper COMM port number and open the port. Turn on the camera. You should receive the "CMUcam V1.12" message. Hit the Dump Frame button and wait. It can take 10 s for the software to dump the frame.

The image shown in the Dump Frame window (see Fig. 14.7) is a simple target I constructed. This target helped me calibrate the camera's field of view. The target is a 2.5-in square of orange paper (see Fig. 14.8), held at a distance of 12 in from the camera lens. I also used this target to read the image processing parameters from my PIC program 2 as I moved the target left, right, up, and down. I assembled these image process readings in a small table; more about this later.

You should use this opportunity to find a good target. Place the object you want to use as your target in front of the camera, and do a frame dump. You are looking to see that the object shows well in the image and has good contrast with the background. You can also see how much space the object takes up in the image. This will give you an idea of how close you should hold the object to the camera.

Once you have your target, you can start using the communication port for issuing commands to the CMU camera. Try turning the green LED on and off

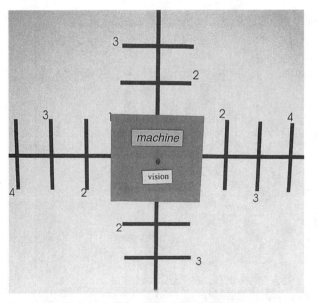

Figure 14.8 Target used for calibrating CMU camera.

as before. You can use this communication port to implement more challenging commands and see the results on the screen dump.

Here are a few commands you may want to try:

Turn on auto light adjustment. This command tells the camera to adjust to the ambient lighting. When you use this command, do *not* have your object/target in front of the camera. The command is `cr 18 44`. Now press the Return key or Send button.

Wait 10 to 20 s for the camera to complete its ambient light adjustment. Then enter this command to *turn off auto light adjustment:* `cr 18 44 19 32`. Now press the Return key or Send button.

This next command I found particularly useful. It turns on a fluorescent band filter with the auto light adjustment: `cr 45 7 18 44`. Now press the Return key or Send button.

You can find other commands in the CMU manual.

Interfacing the CMU Camera to a Robot

The first step in interfacing the camera to a robot is to establish communication between the PIC microcontroller and the CMU camera. Remove the DB9 serial cable used for communicating with the PC. The camera has a TTL serial output next to the jumpers (see Fig. 14.9). Before we can use the TTL serial input/output pins, first we remove the MAX232 IC from the back of the CMU camera.

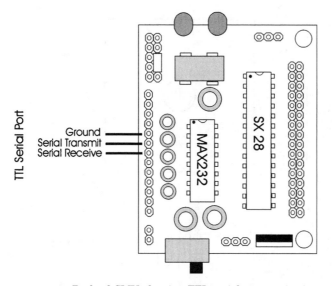

Figure 14.9 Back of CMU showing TTL serial communication jumpers.

Note: At any time you need or want to reconnect the CMU camera serial interface to a PC, you will need to place the MAX232 chip back onto the board.

Plug the TTL cable onto the appropriate header pins on the CMU camera. Figure 14.10 is the schematic we will be using. You do not need to connect the two servomotors for our first two programs.

PIC 16F84 Runs at 16 MHz

One important note about the CMU schematic you must be aware of. The PIC 16F84 used in this circuit is a 20-MHz version operating at 16 MHz with a 16-MHz crystal. I needed to jump up in speed because the 9600-Bd communication is running at the limit of the capacity of 16F84 at 4 MHz. To keep the baud rate timing accurate when we change clock speeds, we enter the command

```
define osc 16
```

This informs the compiler that we are running at 16 MHz. The compiler automatically adjusts the serial commands to keep the baud rate accurate.

Program 1

This first program establishes a communication link between the CMU camera and PIC 16F84 microcontroller. It turns the green LED on the CMU camera on and off. You should not proceed to the more advanced programs until you have this program functioning properly.

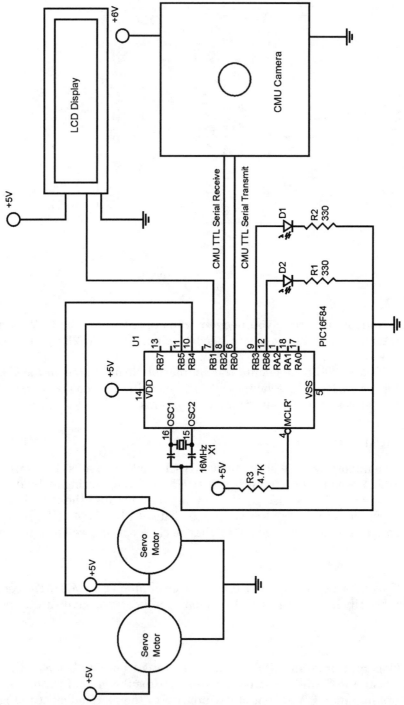

Figure 14.10 Main robot schematic.

When the program starts, it begins with a 5-s countdown. If you look into the countdown loop, you will see that the program issues a reset command each time through the loop. I have found it necessary to send a few reset commands before the camera communication link becomes responsive.

```
'PIC to CMU test
'Send serial information to CMU camera true

define osc 16

x var byte
y var byte
recdata var byte[10]
trisb = 0
portb = 0

pause 1500
serout portb.1,6,["CMU Program V1"]
for x = 0 to 4
y = 5 - x
portb.3 = 1
serout portb.1,6,[254,192,"Starting in",#y]
serout portb.2,2,["RS",13]
pause 500
portb.3 = 0
pause 500
next x

serout portb.1,6,[254,1,"Resetting Cam."]
serout portb.1,6,[254,192]              'Move to second line
serout portb.2,2,["RS",13]
gosub display
pause 1000

start:

'Turn green CMU LED on
serout portb.1,6,[254,1,"Green LED On"]
serout portb.1,6,[254,192]              'Move to second line
serout portb.2,2,["L1 1",13]
gosub display
pause 1000

'Turn green CMU LED off
serout portb.1,6,[254,1,"Green LED Off"]
serout portb.1,6,[254,192]              'Move to second line
serout portb.2,2,["L1 2", 13]
gosub display
pause 1000
```

```
goto start

display:

serin2 portb.0,84,20,error,[str recdata\4]

for x = 0 to 4
serout2 portb.1,16468,[" ",#recdata[x]]
recdata[x] = 32
next x
pause 1000
return

error:
'No acknowledgment
serout portb.1,6,["No ACK - Cont."]
pause 1000
return
```

Program 2

This second program displays on the LCD the major image processing parameters available from the CMU camera. This program just fits into the 1K memory space of the PIC 16F84. If you add a programming line or a couple of letters or spaces in any of the LCD displays, the program will not compile, because it will exceed the PIC 16F84 memory limit. Keep that in mind, if you encounter an error, when compiling this program.

Incandescent or fluorescent lighting

When I first starting working with the CMU camera, I was working under fluorescent lighting. The camera was not tracking its target as well as I expected. Going through the literature I had on the camera, I found a fluorescent filter. I incorporated the filter into my program, and the camera started tracking better. The program uses the fluorescent filter; it is in the following line:

```
'Turn on fluorescent band filter and auto lighting adjust
serout portb.2,2,["CR 45 7 18 44", 13]
```

If you are using fluorescent lighting, you can leave this line alone. However, if your lighting is incandescent, change the command line to

```
serout portb.2,2,["CR 18 44", 13]
```

Obviously this program is more sophisticated than our first program. It displays the type S data packet and then displays the type M data packet in a loop for real-time object tracking. Let's first look at the information that is provided in the type S data packet.

Type S Data Packet

Displayed program parameter	Item	Description
RM	Rmean	The mean red found in the current window
GM	Gmean	The mean green found in the current window
BM	Bmean	The mean blue found in the current window
Rdev	Rdeviation	The deviation of red found in the current window
Gdev	Gdeviation	The deviation of green found in the current window
Bdev	Bdeviation	The deviation of blue found in the current window

Here's a listing of the information that is provided in the type M data packet.

Type M Data Packet

Displayed program parameter	Item	Description
MMX	mx	The middle of mass x value
MMY	my	The middle of mass y value
LCX	x1	The leftmost corner's x value
LCY	y1	The leftmost corner's y value
RCX	x2	The rightmost corner's x value
RCY	y2	The rightmost corner's y value
pix	pixel	Number of pixels in the tracked region
conf	confidence	Number of pixels in area—capped at 255

It's time to choose an object/target if you haven't done so already. Program 2 needs an object to lock onto, too. When the microcontroller runs, it displays information on the LCD screen. During the 10-s autoadjust period, the camera should just be looking at the background. When LED 1 (see schematic) starts to blink, place your object/target in front of the CMU camera.

```
'CMU parameter display program
'By J. Iovine

define osc 16

recdata var byte[10]
x var byte

trisb = 0
portb = 0
```

```
pause 1500
serout portb.2,2,["RS", 13]
serout portb.1,6,["CMU Test Program"]
pause 1000
serout portb.2,2,["RS", 13]
serout portb.1,6,[254,1]

'Reset CMU camera
serout portb.2,2,["RS", 13]
gosub display

'Turn green CMU LED on
serout portb.2,2,["L1 1",13]
gosub display
portb.3 = 1

'Turn on fluorescent band filter & auto lighting adjust
serout portb.2,2,["CR 45 7 18 44", 13]
gosub display

serout portb.1,6,["Auto Adj."]

pause 10000                'Hold 10 seconds

serout portb.1,6,[254,1]
pause 50

'Turn off auto lighting adjust
serout portb.2,2,["CR 18 44 19 32", 13]
gosub display

'Turn green CMU LED off
serout portb.2,2,["L1 2",13]
gosub display
portb.3 = 0

For x = 0 to 10          'Blink red LED to tell user to ready target
portb.3 = 1
pause 250
portb.3 = 0
pause 250
next x

'Set poll mode - 1 packet
serout portb.2,2,["PM 1", 13]
pause 100

'Set raw data
serout portb.2,2,["RM 3", 13]
pause 100
```

```
'Track window command looks at center of CMU window
'Grabs data and sends them to track color function
'Track:
serout portb.2,2,["TW", 13]

'Gather the s statistics packet from TW command
serin2 portb.0,84,[str recdata\8]

'Display data on LCD screen
serout portb.1,6,["RM",#recdata[2]]
gosub hold

serout portb.1,6,["GM",#recdata[3]]
gosub hold

serout portb.1,6,["BM",#recdata[4]]
gosub hold

serout portb.1,6,["RDev.",#recdata[5]]
gosub hold

serout portb.1,6,["GDev.",#recdata[6]]
gosub hold

serout portb.1,6,["BDev.",#recdata[7]]
gosub hold

pause 2000

main:

'Send command - track color (with no arguments)
'Will track last color grabbed by TW command
serout portb.2,2,["TC", 13]

'Gather the m statistics packet from TW command
serin2 portb.0,84,[str recdata\10]

'Display data on LCD screen
serout portb.1,6,["MM-X",#recdata[2]]
gosub hold

serout portb.1,6,["MM-Y",#recdata[3]]
gosub hold

serout portb.1,6,["LC-X",#recdata[4]]
gosub hold

serout portb.1,6,["LC-Y",#recdata[5]]
gosub hold
```

```
serout portb.1,6,["RC-X",#recdata[6]]
gosub hold

serout portb.1,6,["RC-Y",#recdata[7]]
gosub hold

serout portb.1,6,["Pix",#recdata[8]]
gosub hold

serout portb.1,6,["Conf",#recdata[9]]
gosub hold

goto main:

display:

serin2 portb.0,84,20,main,[str recdata\3]

for x = 0 to 3
serout2 portb.1,16468,[" ",recdata[x]]
next x

hold:
pause 500
serout2 portb.1,16468,[254,1]
pause 40
return
```

When the object is captured, the program first displays the S data packet. Then it goes into the main program loop, capturing and displaying the M data packet. Using this program, I constructed a data table that shows how my camera tracked my object/target. In the following table, n/c = no change. Although this isn't 100 percent accurate, I ignored small changes of less than a few points in either direction. The reasons are that (1) I don't want anyone getting bogged down focusing on small changes and missing the important main changes and (2) when I moved a target to the left or right, I didn't keep the height exactly in line. I just moved the target over and kept the height approximately the same. Obviously this caused minor fluctuations that can be ignored.

Data Table ±X (Left and Right)

Parameter	Target 2 in left	Target 1 in left	Target centered	Target 1 in right	Target 2 in right
MMX	67	57	45	35	20
MMY	n/c	n/c	74	n/c	n/c
LCX	53	45	33	23	4

LCY	n/c	n/c	47	n/c	n/c
RCX	80	70	58	53	28
RCY	n/c	n/c	102	n/c	n/c
PIX	144	150	163	162	162
CONF	31	215	232	142	39

From the above table we can make a general observation: As the target moves from left to right, MMX, LCX, and RCX decrease. The reverse is also true; as the target moves to the left, MMX, LCX, and RCX increase.

Data Table ±Y (Up and Down)

Parameter	Target 2 in up	Target 1 in up	Target centered	Target 1 in down	Target 2 in down
MMX	n/c	n/c	45	n/c	n/c
MMY	31	52	74	98	116
LCX	n/c	n/c	33	n/c	n/c
LCY	7	26	47	71	96
RCX	n/c	n/c	58	n/c	n/c
RCY	57	80	102	125	143
PIX	153	157	163	164	116
CONF	246	233	232	237	181

From the above table we can make a general observation. As the target moves up, MMY, LCY, and RCY decrease. The reverse is also true. As the object/target moves down, MMY, LCY, and RCY increase.

Servomotors for robot

This robot we will build uses two HS-425 servomotors modified for continuous rotation. The procedure for modifying these servomotors for continuous rotation was discussed in Chap. 8. Once you have the modified servomotors, it is essential that you determine the pulse widths needed for slow forward, slow backward, and stop.

Figure 14.11 is a schematic for a circuit you can use along with the following PicBasic Pro program to determine the pulse widths. The pulse width is shown in real time on the LCD display. You change the pulse widths up or down by using the SPDT switch. It is essential that the switch used in this circuit have a center-off position.

```
'Continuous rotation servomotor calibration
'Serial communication to LCD display is 2400 baud inverted
x var byte
```

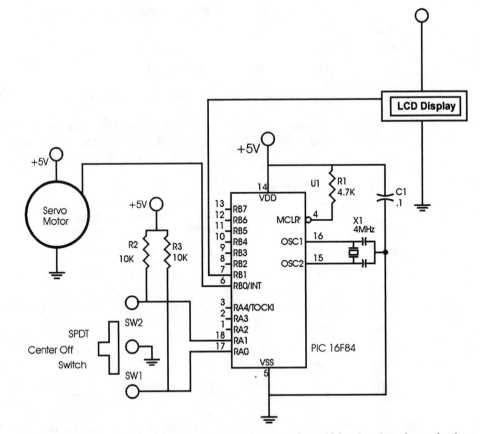

Figure 14.11 Servomotor schematic for determining pulse widths for slow forward, slow backward, and stop.

```
y var byte

pause 1500
serout portb.1,4,["Servomotor Test"]
pause 1000
serout portb.1,4,[254,1]
pause 20
x = 150

main:
pulsout portb.0,x
if porta.1 = 0 then
   x = x + 1
   endif

if porta.0 = 0 then
```

```
x = x - 1
endif

serout portb.0,4,[254,1," ",#x]
goto main
```

Here are the pulse width numbers I needed for the servomotors I used in my prototype robot.

Function	Right servomotor	Left servomotor
Stop	167	169
Slow backward	160	176
Slow forward	174	162

Note that the numbers represent $10\text{-}\mu s$ increments in time. So the 167 used in the program is equal to 1.67 ms.

Program 3

The following program is for our tracking robot. It uses information from the ±X data table to track an object/target from left to right. The PIX pixel parameter is used to determine range of the object. If the object (PIX gets too large) comes too close to the robot, the robot will back away from the object.

Everything stated about program 2 also applies to this program. Keep in mind the lighting—fluorescent or incandescent—and remember to keep the target out of the camera's FOV when it is adjusting for the ambient light. Again this program just fits into the PIC 16F84; so if you add anything to the program, even a few spaces in the display, you stand a good chance of its not compiling properly.

```
'CMU tracking program
'By J. Iovine

define osc 16

recdata var byte[10]
x var byte
confid var byte

trisb = 0
portb = 0

pause 1500
serout port.1,6,["CMU Prg."]
serout portb.2,2,["RS", 13]
pause 1250

'Reset CMU camera
serout portb.2,2,["RS", 13]
```

```
gosub display

'Turn green CMU LED on
serout portb.2,2,["L1 1",13]
gosub display
portb.3 = 1

'Turn on auto lighting adjust & fluorescent band filter ***
serout portb.2,2,["CR 45 7 18 44", 13]
gosub display

serout portb.1,6,["A L"]                     'Auto lighting adjustment

pause 20000                                  'Hold 20 seconds

'Turn off auto lighting adjust
serout portb.2,2,["CR 18 44 19 32", 13]
gosub display

'Turn green CMU LED off
serout portb.2,2,["L1 2",13]
gosub display
portb.3 = 0

'Set poll mode--1 packet
serout portb.2,2,["PM 1", 13]
pause 100

'Set raw data
serout portb.2,2,["RM 3", 13]

for x = 0 to 10              'Blink red LED to tell user to ready target
portb.3 = 1
pause 250
portb.3 = 0
pause 250
next x

portb.6 = 1                                  'Track LED on
'Track window command looks at center of CMU window
'Grabs data and sends it to track color function
'Track:
serout portb.2,2,["TW", 13]

pause 2000

portb.6 = 0                                  'Track LED off

main:
```

```
portb.3 = 1

'Send command--track color (with no arguments)
'Will track last color grabbed by TW command
serout portb.2,2,["TC", 13]

'Gather the m statistics packet from TW command
serin2 portb.0,84,[str recdata\10]

confid = recdata[9]

if recdata[2] > 50 and confid > 20 then left      'MMX
if recdata[2] < 40 and confid > 20 then right     'MMX
if recdata[8] < 175 and confid > 25 then fwd      'PIX
if recdata[8] > 200 and confid > 25 then bwd      'PIX

serout portb.1,6,[254,1,"S"]                      'Stop
   portb.3 = 0
      pulsout portb.4, 668                        'Right servo stop
      pulsout portb.5, 676                        'Left servo stop
      pause 18
   portb.3 = 1
goto main

left:
serout portb.1,6,[254,1,"L", #recdata[2]]
   for x= 1 to 7
      pulsout portb.4, 696                        'Right servo forward
      pulsout portb.5, 676                        'Left servo stop
      pause 20
   next x
goto main:

right:
serout portb.1,6,[254,1,"R",#recdata[2]]
   for x= 1 to 7
      pulsout portb.4, 668                        'Right servo stop
      pulsout portb.5, 648                        'Left servo forward
      pause 20
   next x
goto main:

fwd:
serout portb.1,6,[254,1,"F",#recdata[8]]
   for x= 1 to 7
      pulsout portb.4, 696                        'Right servo forward
      pulsout portb.5, 648                        'Left servo forward
      pause 20
   next
```

```
goto main:

bwd:
serout portb.1,6,[254,1,"B",#recdata[8]]
    for x= 1 to 7
        pulsout portb.4, 640            'Right servo backward
        pulsout portb.5, 704            'Left servo backward
        pause 20
    next x
goto main:

display:

serin2 portb.0,84,20,main,[str recdata\3]

for x = 0 to 3
serout2 portb.1,16468,[" ",recdata[x]]
next x
pause 1500
serout2 portb.1,16468,[254,1]
return
```

Robot construction

By the time this book goes to print, this artificial vision robot will be available as a kit from Images SI Inc. Visit the CMU camera website at http://www.cmu-cam.com. We begin by assembling two part A's of the standard servomotor

Figure 14.12   Two servomotor brackets, part A, assembled.

bracket together (see Fig. 14.12). A front U bracket is made to assemble to the front of the two part A's (see Fig. 14.13). The inside width of the front U bracket is the same as the width of the CMU camera, approximately 2.125 in. The front U bracket has a hole near the front for the shaft of the front wheel. There are holes near the top front of the U bracket also, not shown in the figure.

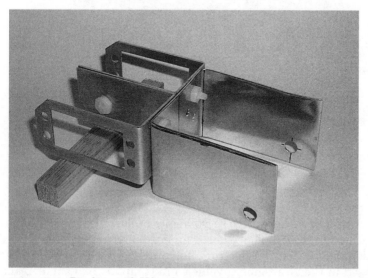

Figure 14.13   Brackets with U bracket assembled.

Figure 14.14   Robot base with servomotors, wheels, and multidirectional front wheel.

Figure 14.15 Finished robot.

Next we assemble our two servomotors and front wheel onto the base assembly (see Fig. 14.14). The wheels for the servomotors are the same type of wheels used in Chap. 8. The front universal multidirectional wheel is the same one used in the Braitenberg vehicles in Chap. 9. Two small L-shaped mounting ears are made to attach the CMU camera to the front of the U bracket.

I constructed the entire circuit on a PIC Experimenter's Board. I changed the Xtal on the board from 4.0 MHz to 16 MHz. Power for the circuit may be obtained from an external power supply or an onboard battery power supply. The finished robot is shown in Fig. 14.15.

Running the Program

When you first run the robot, you may want to have it lifted so the wheels don't touch. It's a lot easier to check operation and function without having to run after the robot. Use the experience you gained with object/targets using program 2. The LED D1 flashes after the auto light adjustment to signal you to put the target in front of the camera.

The D1 LED also flashes when the robot is in the stop loop. I included the flashing LED because it's not always easy to see the LCD display.

The program reads the MMX value from the CMU camera and determines whether the robot should turn left or right. You can adjust these values to suit your particular target. Do not make the greater than (>) and less than (<) values of MMX too close. If you do, the robot will quiver left and right constantly.

If you find the robot constantly overshooting when it turns to the left or right, you can reduce the loop value (x) in these turn subroutines.

Going Further

Obviously, we have just scratched the surface of playing with the CMU camera. One feature I didn't have time to implement was an up-and-down tilt servomotor that uses the MMY parameter. This involves adding another servomotor to the robot, but would allow the robot to follow a target as it moves up and down.

I quickly approached the memory limit of the PIC 16F84. If I had had more time, I would have implemented it, using another PIC microcontroller with a little more memory. The PIC 16F628 is port B–compatible with the 16F84 and has twice as much memory (2048 bytes).

Latest updates and information on the CMU camera can be found at http://www.cmucam.com or http://www.cmucamera.com.

Parts List

CMU camera

(2) Servomotors (HS-425)

(2) Part A servomotor brackets

16-MHz PIC Experimenter's Board

(2) Servomotor wheels

PIC 16F84, 20 MHz

16-MHz crystal

(2) 22-pF capacitors

(2) 330-Ω, $\frac{1}{4}$-W resistors

4.7-kΩ, $\frac{1}{4}$-W resistor

Multidirectional wheel

Aluminum sheet metal, shaft, plastic screws, and nuts

Available from Images SI Inc. (see Suppliers at end of book).

Suppliers

Images SI Inc.
109 Woods of Arden Road
Staten Island, NY 10312
(718) 698-8305
(718) 982-6145 (fax)
www.imagesco.com

Jameco Electronics
1355 Shoreway Road
Belmont, CA 94002
(800) 831-4242
(800) 237-6948 (fax)
www.jameco.com

JDR Microdevices
1850 South 10th Street
San Jose, CA 95112
(800) 538-5000
(800) 538-5005 (fax)
www.jdr.com

Index

ABOUT THE AUTHOR

John Iovine is the author of several popular TAB titles that explore the frontiers of scientific research. He has written *Homemade Holograms: The Complete Guide to Inexpensive, Do-It-Yourself Holography; Robots, Androids, and Animatrons: 12 Incredible Projects You Can Build,* considered a cult classic; *Kirlian Photography: A Hands-On Guide; Fantastic Electronics: Build Your Own Negative-Ion Generator and Other Projects;* and *A Step into Virtual Reality*. Mr. Iovine has also written extensively for *Popular Electronics, Nuts & Volts, Electronics Now,* and other periodicals.